PEDESTRIANS, URBAN SPACES AND HEALTH

PROCEEDINGS OF THE XXIVTH INTERNATIONAL CONFERENCE ON LIVING AND WALKING IN CITIES (LWC 2019), BRESCIA, ITALY, 12–13 SEPTEMBER 2019

Pedestrians, Urban Spaces and Health

Editors

Maurizio Tira, Michèle Pezzagno & Anna Richiedei

University of Brescia, Italy

CRC Press
Taylor & Francis Group
Boca Raton   London   New York

CRC Press is an imprint of the
Taylor & Francis Group, an **informa** business

A BALKEMA BOOK

CRC Press/Balkema is an imprint of the Taylor & Francis Group, an informa business

© 2020 Taylor & Francis Group, London, UK

Typeset by Integra Software Services Pvt. Ltd., Pondicherry, India

Library of Congress Cataloging-in-Publication Data

Applied for

Published by: CRC Press/Balkema
　　　　　　 Schipholweg 107C, 2316XC Leiden, The Netherlands
　　　　　　 e-mail: Pub.NL@taylorandfrancis.com
　　　　　　 www.routledge.com – www.taylorandfrancis.com

ISBN: 978-0-367-46171-3 (Hbk)
ISBN: 978-1-003-02737-9 (eBook)
DOI: 10.1201/9781003027379
https://doi.org/10.1201/9781003027379

Pedestrians, Urban Spaces and Health – Tira, Pezzagno & Richiedei (eds)
© 2020 Taylor & Francis Group, London, ISBN 978-0-367-46171-3

Table of contents

Pleasant and attractive public spaces

Soft mobility and perception of urban landscape

Pedestrian road safety

Preface

The XXIV International Conference *Living and Walking in Cities (LWC)* was organized under the auspices of the University of Brescia. The Conference was organized with the patronage of: Region of Lombardy, Italian National Institute of Town Planning (INU), Italian Society of Urban planners (SIU), National Centre for Urban Studies (CeNSU), European Transport Safety Council (ETSC), Italian Association for Traffic and Transportation Engineering, Province of Brescia, Municipality of Brescia, Brescia Local Public Transport Agency; and with grants from: Brescia Mobilità and Brescia Infrastrutture.

The Conference was held September 12-13, 2019 in Brescia, Italy.

The Conference traditionally looks at different themes concerning the quality of life in urban areas. The goal of this event was to gather researchers, road users, administrators, technicians, city representatives and experts aiming to discuss problems that affect the safety of pedestrians in the city, especially of children and persons with reduced mobility. The Conference attracted practitioners and researchers who could find detailed presentations on policy issues, best practices and research findings across the broad spectrum of urban and transport planning. The Conference covered international issues, national and local policies and the implementation of projects at the local level.

The Conference presented a great opportunity for networking and forming career-spanning professional relationships. Although sessions at the Conference could be challenging in discussing matters of policy at the highest level, they could also provide good basic education and training opportunities. The Conference *Living and Walking in Cities (LWC)* provided a forum to discuss the challenges faced by economic growth, social and demographic changes in order to become more sustainable. Planners and practitioners were being asked to improve and retrofit towns, transportation infrastructure and public spaces; they were searching solutions to build resilience in the face of threats posed by climate change and by the need to improve energy and infrastructure security. At the same time, they needed to develop hard and soft measures to improve the safety for walking and cycling, which affects health and fitness.

The topic selected for the 2019 edition was "*Pedestrians, Urban Spaces and Health*" aiming at linking the traditional topic of walkability with the challenges caused by an ageing population and a widespread demand for health.

The Conference focused on a wide spectrum of topics listed below:

- Networks and infrastructures to improve pedestrian mobility
- Urban space for non-motorized traffic and public transport
- Soft mobility and perception of the urban landscape
- Walking, health and wellbeing
- Sustainable and resilient urban spaces
- Safety of urban environment
- Public space and beauty
- Urban leisure environment
- Urban environment and health
- Urban form and heath
- Open spaces and green infrastructures

- Promoting healthy cities
- Healthy city for all
- The role of public space in attracting investments

Forty-six papers, whose authors came from fourteen countries and thirty-eight different universities and five research institutes, were accepted for publication in the proceedings of the Conference. Each of the accepted papers was reviewed by selected members of the Scientific Committee in accordance with the scientific area and orientation of the paper itself.

The editors would like to express their many thanks to the members of the Organizing and the Scientific Committee. The editors would also like to express a special thanks to all reviewers, sponsor and Conference participants for their intensive cooperation to make this Conference successful.

Maurizio Tira

Michèle Pezzagno

Anna Richiedei

Organizing Committee

Conference Chairman:
Maurizio Tira - *Dipartimento DICATAM. University of Brescia (IT)*

Emeritus Conference Chairman:
Roberto Busi - *Dipartimento DICATAM. University of Brescia (IT)*

Scientific Committee:
Richard Allsop - *Department of Civil, Environmental and Geomatic Engineering, Faculty of Engineering Science, UCL University (UK)*
Deodato Assanelli - *Dipartimento SCS. University of Brescia (IT)*
Antonio Avenoso - *European Transport Safety Council (BE)*
Ivan Blečić - *Dipartimento DICAA. University of Cagliari (IT)*
Roberto Busi - *Dipartimento DICATAM. University of Brescia (IT)*
Enrique J. Calderon - *ETS de Ingenieros de Caminos, Canales y Puertos. The Technical University of Madrid (ES)*
Mattero Ignaccolo - *Dipartimento Ingegneria Civile e Ambientale. University of Catania (IT)*
Giulio Maternini - *Dipartimento DICATAM. University of Brescia (IT)*
Claudio Orizio - *Dipartimento SCS. University of Brescia (IT)*
Graham Parkhurst - *FET Geography and Environmental Management. University of the West of England (UK)*
Michele Pezzagno - *Dipartimento DICATAM. University of Brescia (IT)*
Tim Pharoah - *Transport and urban planning consultant (UK)*
Paola Pucci - *Dipartimento di Architettura e Studi Urbani. Politecnico Milano (IT)*
Michela Tiboni - *Dipartimento DICATAM. University of Brescia (IT)*
Maurizio Tira - *Dipartimento DICATAM. University of Brescia (IT)*
Rodney Tolley - *Sustainable Transport Consultant (UK)*
Paolo Ventura - *Dipartimento DIA. University of Parma (IT)*
Michele Zazzi - *Dipartimento DIA. University of Parma (IT)*

Scientific Secretariat:
Anna Richiedei - *Dipartimento DICATAM. University of Brescia (IT)*

Web site of the conference: http://lwc.unibs.it/

Published with the contribution of: University of Brescia, Brescia Infrastrutture and Brescia Mobilità

Introduction

Pedestrians, Urban Spaces and Health – Tira, Pezzagno & Richiedei (eds)
© 2020 Taylor & Francis Group, London, ISBN 978-0-367-46171-3

Pedestrians, urban spaces and health

M. Tira
University of Brescia, Italy

Ten years ago, I had the great opportunity to join the working group of the International Transport Forum-OECD, composed of experts from 19 countries as well as from the World Health Organisation. The report produced in two years work is entitled Pedestrian safety, urban space and health.

The report assembles the most relevant informations on the importance of walking to the future development of our cities, including the benefits of walking for health and wellbeing.

At the end of the experts' report, the need for a walking strategy, including the role that governments and stakeholders can play, raised. Seven years after the publication, that is still a major challenge for urbanists and mobility planners.

The 2019 Conference recalls the title of that report and focuses on the implications of the problem, updating our scientific evidence after so many research evidences. The organisation of the conference is on three blocks of topics.

The infrastructure related issues

The research group of the University of Brescia belongs to an engineering Faculty, so the interest for network design and dedicated infrastructure to improve pedestrian mobility is of utmost importance. The main research questions are related to the obstacles to the implementation of relatively simple measures: lack of funding, un-coordinated competencies, poor urban design, missing clear political priorities. So many guidelines have been produced to describe forgiving roads, shared spaces, sustainable pedestrian and cycling networks, but the rate of realization is still largely unsatisfactory.

We could also mention a renovated interest for sustainable and resilient urban spaces, that also require re-design of several parts of urban settlements. Resilience is the new name for describing the ability to face risk and adapt urban shape to external inputs. So risk theory is at stake when approaching a safe urban design, including actions against natural risks and transport accidents.

The safety related issues

The research tradition of our group is well routed into the accident analysis to propose technical solutions in order to increase urban safety. The process starts with planning in order to enhance road safety and continues with road and urban design. As far as sustainability is a shared goal, safety becomes more and more relevant to urban governance.

Vulnerable road users' safety is particularly challenging as the protection of the weakest is the lever to implement a broader set of safety measures for all. The relatively poor performance of road safety measures and the new transport modes and related new conflicts keep interest on these topics alive.

Pedestrian movement and the health of people

The last perspective of the conference is the link between pedestrian movement and the health of people, as the latest findings of medical research clearly confirm the growth of medical expenses related to bad habits and sedentary lifestyle.

A growing medical care expense is unsustainable for the public bodies and a growing mortgage is poured on the shoulders of the next generations.

Among public policies those related to health promotion seem to be more promising, so that we can foster the combined effects of improving physical activities, reducing motorized traffic and increasing liveability.

The general perspective under which we run the Living and Walking in Cities conferences is the link between urban space (either planned or not) and movement.

The decrease in walking trends are related to urban planning. It seems, for example, that the acceptable walkable distance is increasing with the size of the core city. On the other side, in small and less densely populated villages, especially when no facilities are provided, people accept less the long trips.

When recalling the relation speed/distance, applicable also to pedestrian movement, it is clear that the low speed (and consequently the longer time needed) of pedestrian movement influences walking trip length and the number of trips as well. So controlling the space/time relation in planning is crucial to restore an higher level of pedestrian mobility.

A relevant reason why walking less is the lost proximity: people walk less because there are few destinations within a walkable distance. The traditional zoning separating functions determines longer average trips, shopping malls can be reached only by car, services are concentrated for economic reasons and work places are not fixed, so trips are multi-scope and they need flexible means of transport.

New scenario must be designed to invert this trends.

Therefore, through land use planning and urban design it is possible to highly improve conditions for pedestrian movement (to create an accessible, comfortable, safe and attractive environment) but also pedestrian safety (by improving safety conditions in the urban environment).

When looking at town planning, three main phenomena occur:

- the building of city extensions, consuming new land but easier for implementing mobility networks and also pedestrians' friendly schemes;
- the reconstruction of cities, through brown-field regeneration, that must take into account the relation between administrators and developers;
- the new implementation of transport network in the existing urban structure.

Most of the countries are facing these phenomena, with different proportion of examples in the three cases.

Among others, the "strategy setting" and 'sharing interests' turned out to be successful safety policies, in those European Countries where applied. It is an approach that looks at urban environment from the point of view of global safety and comfort, pointing them as the core strategy for any action. Setting targets is part of the strategy.

The target specifically set by the United Nations' Sustainable Development Goals for transport by 2030 is: "provide access to safe, affordable, accessible and sustainable transport systems for all, improving road safety, notably by expanding public transport, with special attention to the needs of those in vulnerable situations, women, children, persons with disabilities and older persons and provide universal access to safe, inclusive and accessible, green and public spaces." (UNDP, 2015).

The main goal of the Living and Walking in City Conferences is exactly that: improving urban quality while promoting safer mobility, especially for the most vulnerable road users. The way vulnerability is tackled is one of the most important indicators of civilisation and the "design for all" approach appears to be the most promising one.

Just as housing demand has been the lever of urban development, involving a great portion of actual urban public space, mobility management and road safety can play the role of a lever for improving town management and townscape as well. For this reason this topic is relevant for urban planners, not just transportation and traffic engineers.

Even the most traditional approach to road safety, i.e. that of treating 'clustered' accidents, often hides the inner relation between town structure and road (un)safety. The so-called 'most dangerous spots' represent just the tip of the iceberg that can be depicted as a generalised problem of friendly and safe mobility.

There is now a considerable effort to combine the planning of land use and transport in many European countries, although this is mainly because environmental problems caused by private motorized transport in the form of car-emitted gases, noise, water and land pollution, have become crucial issues. However, this situation can often be used as a catalyst for change and it may lead to a reduction in car trips that can benefit road safety and vulnerable road user comfort.

Focusing on pedestrians is not just a matter of lobbying for their own benefit, rather a better mix of modes can be an important factor in improving safety for all. However, modal shift is a difficult challenge, and the lack of safety imposed by cars is paradoxically one of the reasons for using cars even more.

The results of the international debate about Pedestrians, urban spaces and health were discussed and developed during the International Conference "Living and Walking in Cities" (LWC, 2019) that took place in Brescia on September 2019 and its highlights are reported in this book, thanks to the great job done by the Conference organization, namely Michele Pezzagno and Anna Richiedei.

Network and infrastructure to improve

pedestrian mobility

Pedestrians, Urban Spaces and Health – Tira, Pezzagno & Richiedei (eds)
© *2020 Taylor & Francis Group, London, ISBN 978-0-367-46171-3*

E-thinking the road infrastructure for new urban mobility needs

G. Cantisani
Department of Civil, Construction and Environmental Engineering, Sapienza University of Rome, Rome, Italy

ABSTRACT: Urban road infrastructure, for a long time, had been designed and built on the basis of theoretical and practical knowledge, focused on the urban functions development and motorized traffic requirements. As a matter of fact, vulnerable users' functions were considered as an optional or a secondary objective.

Nowadays, a different paradigm of urban mobility is establishing, with particular attention to the safety and environmental performances of infrastructures. The main subject to propose is to overcome a standard-based design in favour of a needs-driven design, with the aim to balance the conflicting characteristics and expectations of the various traffic components. Some examples can be proposed, at various level of scale and/or for typical infrastructure.

Various situations are presented, in the paper, with the aim to discuss how the infrastructures could be reconsidered and redefined, in order to become safer and more effective for all the traffic components.

1 INTRODUCTION

The current features of urban road infrastructure generally are the result of historical evolution processes and previous planning actions, that rarely had considered – in detail – the problem of vulnerable users.

The general approach to road design and construction, in the past, was essentially based on a model focused on the urban functions development and motorized traffic requirements [Hawbaker et al., 2006; Rui et al., 2011]; the needs of pedestrians and cyclists have been considered, for a long time, as an optional or a secondary objective.

The result, also from the point of view of road researchers and designers, present various and serious problems in order to face the real needs of pedestrians, cyclists and other vulnerable users, especially with regard of safety problems that are increasing along the road networks in urban areas [CARE, 2016; Osservatorio Utenze Deboli, 2011].

It is important to carefully consider these problems and to develop a retrospective evaluation of the processes that determined the current conditions of road infrastructures, with the aim to correct previous mistakes and to orient future actions aimed to design or re-design more sustainable road infrastructures.

2 ROAD NETWORKS CHARACTERISTICS

2.1 *The historical evolution of road infrastructure*

The characteristics of road infrastructure, in a certain sense, have changed in a very tumultuous way in a relatively short time.

In the 20th century, in fact, after more than seven thousand years [Da Rios, 2010] during which roads had served a different kind of mobility (Figure 1), the mass motorization completely changed the model and changed the concept of the road infrastructure itself [Cervero, 2003].

Figure 1. Before the mass motorization, for many centuries roads had served to a "natural" mobility model (in the pictures: the excavation of a section of the ancient Via Laurentina in Rome, 2011).

Since then, the general approach to the road design and construction has been essentially based on a model focused on the urban expansion and the correspondance to traffic volumes and motorized vehicles' needs. The growth of cities has been focused on vehicular traffic flows, on a structure often determined by the historical heritage (Figure 2).

Figure 2. In the growth of cities the design of roads has been essentially aimed to the vehicular traffic flows requirements (in the picture: Ernst-Reuter-Platz in face of the Technische Universität Institut, Berlin).

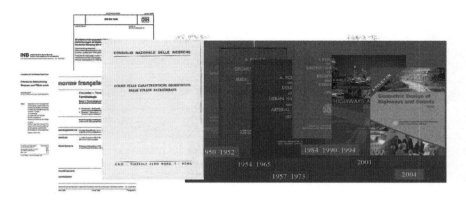

Figure 3.   Technical standards often have driven the design and construction of road infrastructure.

Roads – and not just highways - were conceived as complementary and finalized systems for the massive transit of vehicles [Topp, 1989; Pécheux, 2004]. The needs of pedestrians, cyclists and other vulnerable users have been considered, for many decades, as an optional or a secondary target.

Geometrical and structural features – like the width of lanes, the composition of carriageway, the alignment, the longitudinal slopes, the bearing, the pavement... – evolved and changed to match characteristics and physics of motorized vehicles.

As a result, the road network assets are currently composed of elements and facilities, especially conceived for heavy traffic flows; they do not allow to easily satisfy other users' mobility and safety demand.

2.2   The role of design standards

During the described process, road design and construction were defined – almost all over the world – by standards and technical regulations [for example: DM-IT, 2001; RAS-L, 1995; AASHTO, 2001] that define road characteristics, dimensions, composition and structure as a function of road category (type), traffic flow characteristics and required operative/design speed (Figure 3). This policy was actually proper, on the one hand, in order to achieve uniformity and standardization of road infrastructure everywhere, with consequent improvement of the safety conditions for motorized traffic flows.

But, unfortunately, this trend involved both rural and urban areas; the standards often neglected or not considered the special features of urban environment, especially because the most sensitive variables for motorized traffic (speed and vehicle dimensions) generally are smaller in this context.

As a matter of fact, after some decades in which this regulation had been experimented, other critical issues have arisen: pollution, congestion, occupation of urban spaces and, above all, an insidious problem of road safety, especially due to the high concentration of vulnerable users.

3   A NOVEL APPROACH TO URBAN INFRASTRUCTURE DESIGN

After the historical processes, summarized above, a different awareness of the complex problems that mobility involves in urban areas is growing. In particular, the need to improve sustainability in terms of environmental, economical and social topics, is promoting a different mobility model, in which vehicular traffic is well controlled and pedestrians and cyclists can have more safety and quality.

For example, the release of the Urban Street Design Guide by the National Association of City Transportation Officials (NACTO) provides new engineering standards specifically designed for cities [NATCO, 2013]. As said by Dylan Reid of the University of Toronto [Reid, 2013], these standards «...*are meant to not only ensure the safety of all city users, including pedestrians, cyclists and transit users as well as drivers, but also create cities that are vibrant and appealing*». On these basis: «*Cities are leading the movement to redesign and reinvest in our streets as cherished public spaces for people, as well as critical arteries for traffic...*».

Other authorities, researchers and professionals are presenting their proposals; so, as a general remark, we can say that a different paradigm of urban mobility is establishing, and specific targets are proposed to the attention of urban planners and road engineers, with particular regard to the safety performances of infrastructures, especially for vulnerable users.

Probably, the main subject to propose is to overcome a standard-based design in favour of a needs-driven design, that considers and analyses the real urban mobility demand and its special features, with the aim to better balance the conflicting characteristics and expectations of the various traffic components.

Therefore, not only more advanced standards for plans and design are needed, but also a different approach to the topics of urban mobility and transportation facilities. This requires a big effort and a deep change of mind by the main institutional subjects that have in charge the development of urban systems.

4 EXAMPLES OF NEW INFRASTRUCTURE

To better clarify the possible novel approach to the urban mobility infrastructure design, some examples is proposed, in the following, for various level of scale and/or for some typical infrastructure.

4.1 *Road network structure and classification*

Traditionally, the road network classification is based on a rigid structure that distinguishes four hierarchical levels: primary, principal, secondary and local [AASHTO, 2001; Forbes, 1999]. In the design standards, road types and their features are referred to these levels.

However, frequent violations of this structure and increasing situations in which "hybrid" roads are accepted – also as new construction – highlight the need to better understand the functions and characteristics of these urban infrastructures. For example, in many cases the continuity of traffic flows along principal arterials is frequently interrupted by pedestrian crossings, at-grade intersections, traffic lights. Other problems concern the legal or illegal parking and consequent vehicular manoeuvres on the lateral lanes, the promiscuity of too different kinds of vehicles, the occupation of bus lanes and so on.

These facts ask us if a more complex and well-structured method for road classification is needed, in order to recognize the most suitable technical features to provide for each road segment and road element.

In this sense, new criteria have been proposed, for example, by the *"Seattle right-of-way improvements manual"* [City of Seattle, 2019], as showed in the following matrix (Figure 4).

This approach could be a good example for the revision, also in the European context, the traditional road networks' structure.

4.2 *Street elements*

A wide field of action where new design concept may be proposed is the composition amd the design definition of urban streets. A lot of types of street elements (sidewalks, travel lanes, transit stops, curbs, ...) have to be selected and coordinated in the road design. These street elements can have different characteristics and can be arranged with various criteria. These

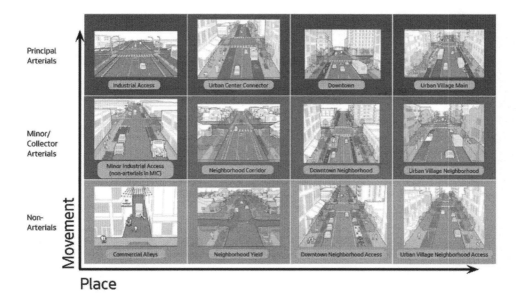

Figure 4. Street type standards according to the *"Seattle right-of-way improvements manual"* [City of Seattle, 2019].

Figure 5. Examples of street elements aimed to traffic calming effect.

design actions become especially important if they are also aimed to obtain vehicular traffic calming effect (Figure 5) [Maternini et al. 2010]. In this case, in fact, the design should study in detail the users' response to the presence of the street elements and consider if these behaviours agree with the desired ones [Maternini, 2017].

13

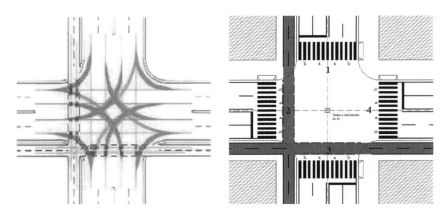

Figure 6. The accurate analysis of trajectories and conflicts is required for the design of urban road intersections, as well as the sharing of space between different kind of users.

As a general method, therefore, for each situation an accurate and attentive design is required, as well as a subsequent monitoring action because it is possible that successive modifications and regulations can occur and can modify the original traffic conditions.

4.3 *Road intersections*

Road intersections are a primary topic in order to improve the coexistence between different kind of users in urban areas (motorized vehicles, bicyclists, pedestrians, ...) and especially to prevent road conflicts, accidents, fatalities, injuries, and so on. In this context, in fact, vehicles and users, characterized by different manoeuvrability and various levels of vulnerability, have to share the same space; this often involves conflicts and accidents [Cantisani et al., 2019; Li et al. 2015]. In addition, other externalities like pollution, noise, space occupation, can also become critical in the urban intersection areas.

Therefore, the deepen study of the design problems of road intersections (Figure 6) should include many issues [Mauro, 2015; Cantisani et al., 2019] like, for example: the analysis of trajectories, the resolution (or mitigation) of conflict points, the assessment of the level of service, the regulation of flows by means of traffic lights and so on.

Also for the case of roundabouts [Sakshaug et al., 2010; Fortuijn et al., 2003], new solutions have to be addressed, because the position of pedestrian and cycling crossing and the overextension of their paths are a common problem and induce inappropriate behaviours of users.

5 CONCLUSIONS

After a long time in which the growth of cities had been focused on vehicular traffic flows, on a structure often determined by the historical heritage, the needs of pedestrians, cyclists and other vulnerable users cannot to be neglected any more.

Safety problems in urban areas are increasing, as well as pollution, congestion, occupation of spaces and other various externalities; these conditions require new solution for the design and re-design of road infrastructure.

Nowadays, a different awareness of the complex problems that mobility involves in urban areas is establishing. In particular, the need to improve sustainability in terms of environmental, economical and social issues, is promoting different mobility models, in which vehicular traffic is well controlled and pedestrians and cyclists can be safer and more satisfied.

More general, a different paradigm is establishing, and this challenge will change previous habits of planners and technicians. Urban planners, together with road engineers, are ingaged in this difficult task.

In this paper, some example of the new possible approach to the design problems have been proposed: guidelines, proposals and updated design manuals can be presented as the evidences of this new trend.

If the focus is addressed on road infrastructures, specific problems can be highlighted, as examples at various level of scale and/or for some typical infrastructures. In particular, the problems related to: road network classification, design of street elements, design and regulation of road intersections, have been presented, pointing out the differences between the traditional way to face these technical tasks and possibile new approaches to define them.

More general, urban planners and road engineers are required to propose a novel urban mobility paradigm, that allow to achieve specific targets in terms of sustainability of road infrastructure and their exercise, with particular regard to the safety performances (especially for vulnerable users) and environmental conditions.

REFERENCES

AASHTO American Association of State Highway and Transportation Officials. (2001). A policy on geometric design of highways and streets, *2001. Washington, D.C: American Association of State Highway and Transportation Officials*.

Cantisani, G., Moretti, L., & De Andrade Barbosa, Y. (2019). Safety problems in urban cycling mobility: a quantitative risk analysis at urban intersections. *Safety, 5 (1)*, 6.

Cantisani, G., Moretti, L., & De Andrade Barbosa, Y. (2019). Risk Analysis and Safer Layout Design Solutions for Bicycles in Four-Leg Urban Intersections. *Safety, 5 (2)*,24.

City of Seattle (2019). Seattle right-of-way improvements manual. https://streetsillustrated.seattle.gov/street-types/(accessed: *Aug, 2019*).

Cervero, R. (2003). Road expansion, urban growth, and induced travel: A path analysis. *Journal of the American Planning Association, 69(2)*, 145-163.

Community Road Accident Database (CARE). (2016). Community Road Accident Database; *CARE: New York, NY, USA, 2016*.

Da Rios, G. (2010). Settemila anni di strade. *EDICEM*.

DM-IT - Ministero delle Infrastrutture e dei Trasporti. Norme Funzionali e Geometriche per la Costruzione delle Strade. *Ministero delle Infrastrutture e dei Trasporti: Rome, Italy, 2001*.

Forbes, G. (1999). Urban roadway classification. *In Urban Street Symposium*.

Fortuijn, L. G. H. (2003). Pedestrian and Bicycle-Friendly Roundabouts; Dilemma of Comfort and Safety. *In Annual Meeting of the Institute of Transportation Engineers (ITE), Seattle, Washington (USA)*.

Hawbaker, T. J., Radeloff, V. C., Clayton, M. K., Hammer, R. B., & Gonzalez-Abraham, C. E. (2006). Road development, housing growth, and landscape fragmentation in northern Wisconsin: 1937–1999. *Ecological Applications, 16(3)*, 1222-1237.

Li, B.; Xiong, S.; Li, X.; Liu, M.; Zhang, X. (2015). The Behavior Analysis of Pedestrian-cyclist Interaction at Non-signalized Intersection on Campus: Conflict and Interference. *Procedia Manuf. 2015, 3*, 3345–3352.

Maternini, G., & Foini, S. (2010). Tecniche di moderazione del traffico. Linee guida per l'applicazione in Italia. *XIV Volume della collana Tecniche per la sicurezza in ambito urbano, EGAF; Forlì, 2010*.

Maternini, G. (2017). The Bristol Method: How to reduce traffic and its impacts. *CSE-City Safety Energy, (1)*, 73-75.

Mauro, R. (2015). Traffic and Random Processes; *Springer International Publishing: Cham, Switzerland, 2015*.

NATCO, (2013). Urban Street Design Guide - 3rd Edition. *National Association of City Transportation Officials. ISBN: 9781610914949*.

Osservatorio Utenze Deboli. L'insicurezza stradale. *In Incidentalità Urbana; Osservatorio Utenze Deboli: Rome, Italy, 2011*.

Pécheux, K. K., Flannery, A., Wochinger, K., Rephlo, J., & Lappin, J. (2004). Automobile drivers' perceptions of service quality on urban streets. *Transportation research record, 1883(1)*, 167-175.

RAS-L - Forschungsgesellschaft für Strassen-und Verkehrswesen; Richt-linien für die Anlage von Strassen: Linienführung RAS-L. *Bonn, 1995*.

Reid, D. (2013). Re-engineering road standards for cities. http://spacing.ca/toronto/2013/09/24/reid-re-engineering-road-standards-cities/

Rui, Y., & Ban, Y. (2011)Urban growth modeling with road network expansion and land use development. *In Advances in Cartography and GIScience. Volume 2 (pp.* 399-412)*Springer, Berlin, Heidelberg.*

Sakshaug, L., Laureshyn, A., Svensson, Å., & Hydén, C. (2010)Cyclists in roundabouts—Different design solutions. *Accident Analysis & Prevention, 42(4)*, 1338-1351.

Topp, H. H. (1989)Traffic safety, usability and streetscape effects of new design principles for major urban roads. *Transportation, 16(4)*, 297-310.

Pedestrians, Urban Spaces and Health – Tira, Pezzagno & Richiedei (eds)
© 2020 Taylor & Francis Group, London, ISBN 978-0-367-46171-3

Sensitivity analysis and the alternative optimization of the pedestrian level of service: Some considerations applied to a pedestrian street in Greece

T. Campisi, A. Canale & G. Tesoriere
Faculty of Engineering and Architecture Cittadella Universitaria Enna (EN), University of Enna KORE, Italy

S. Basbas, A. Nikiforiadis & P. Vaitsis
Department of Transportation & Hydraulic Engineering, School of Rural & Surveying Engineering, Aristotle University of Thessaloniki, Thessaloniki, Greece

ABSTRACT: Pedestrian and cyclable mobility can offer environmental, social and economic benefits to the cities. These two modes of transport belonging to sustainable transport alternatives. Policymakers promote these means of transport in order to reduce the emissions of private's car usage. The growing demand for pedestrian and cyclable mobility requires the policymakers to make interventions to some infrastructures, especially in urban areas. Such interventions can be assessed either ex-post or ex-ante through the examination of the Level of Service (LOS). Therefore critical points can be intentified in the event of congestion. In the framework of the present paper a comparison of scenarios referrred to mixed traffic in an area located in the Municipality of Kalamaria, Greece is made.The interaction between the various modes of road traffic is taken into account in the analysis

1 INTRODUCTION

In many countries, the increase of private car usage leads to significant social, environmental and economic problems. Therefore, the development of non-motorised transport systems has been taken into consideration. In addition, if in the twentieth century the development and vitality criterion of a city was the wide-open public green space and wide appropriate footpaths, in the present century areas and urban footpath networks are one of the most important spaces of leisure and the obvious evidence of the use of engineering design knowledge in the process of the development of cities. Any activity that took place in the city together with its physical system requires adequate decision making.

A sensitivity analysis can evaluate not only the impact of outliers but also the impact of protocol deviations in randomized clinical trials and the impact of the missing data that are common in every research study. A possible strategy is to evaluate the data by excluding the missing values, thus monitoring only the complete data.By imputing the missing values through single or multiple imputations and then bringing them back data were analysed. The examination of capacity in urban footpaths at administrative and commercial districts by the management authorities received special attention during the recent years. Thus, the efficient facilitation of pedestrian flows which imposes increase of the PLOS interventions, cost reduction and economic growth can be achieved. PLOS defined in this paper has considerable difference with PLOS mentioned in the US Highway Capacity Manual (TRB, 2016) due to the highly heterogeneous traffic flow on main roads, poor implementation of traffic rules, the difference in road geometry, activities of unauthorized vendors, unwanted barrier utilities and off-set.

2   OBJECTIVE OF THE PAPER

The daily problems of pedestrians, who are considered as vulnerable users, can be mainly investigated based on the experience acquired by the pedestrian who moves periodically in a specific area. Besides, the individual gait is also connected to the problem of congestion as well as to comfort, safety, and environmental context. In urban areas, studies on slow mobility are carried out in analogy with the vehicular outflow, considering quantitative measures such as the freedom to maintain the desired pedestrian speed and the possibility of interrupting other pedestrians.

The study includes an initial analysis of the existing situation as an observed scenario and subsequently two other hypothetical scenarios which were compared through microsimulation, maintaining constant vehicular and cycling flows and compositions. The variable which presents a different value in the 3 scenarios is the pedestrian flow.

The correlation of these variables as well as provision of an overall judgment is made feasible through the evaluation of the PLOS. This allows for the definition of the possibility given to pedestrians for using the same infrastructure without problems or with limited number of problems in accordance with (Nikiforiadis & Basbas, 2019).

The micro simulation software PTV Viswalk (PTV Group, 2011) linked to the Helbing model theory (Helbing & Molnar, 1995) is used to calculate LOS.

The purpose of this paper is to investigate the Level of Service (LOS) of a street called Metamorfoseos which is located in Kalamaria, Greece. This study allows a comparison of different scenarios of cycle-pedestrian traffic. In agreement with (Panagopoulos, Tampakis, Karanikola, Karipidou-Kanari, & Kantartzis, 2018) the safety and convenience of the mobility of residents is among the most important advantages of pedestrian streets.

3   METHODOLOGICAL APPROACH AND/OR RESEARCH APPROACH

Sensitivity analysis is a method to determine the robustness of an assessment by examining how much of the results can be influenced by changes in methods, models, unmeasured variable values or hypothesis.This type of analysis is focused on the identification of results which depend largely on questionable or unsupported hypotheses.

Fruin (Fruin, 1971) observed pedestrian flow characteristics based on macroscopic approach, which was adopted by TRB in 1985. Pedestrian flow characteristics can be analysed using three parameters, namely speed (U), flow (Q) and density (k). Therefore, sensitivity approach defines an important role in verifying the robustness of a study's conclusions. If the results remain robust under different assumptions, methods or scenarios, this can strengthen their credibility.

The vehicle, bicycle and pedestrian flows were monitored during peak hours.Speed is measured and evaluated through the use video cameras and software. These values were used to calibrate the micro-simulation tool in order to be able to evaluate the LOS, the pedestrian/cyclist speed and the density as the observed and hypothesized scenarios change.

It was then possible to implement the Origin-Destination(O/D) matrices for different scenarios with a mix of cycle-pedestrian traffic. The PTV Vissim software (PTV Group, 2015) allows for the LOS estimation in the study area in accordance with the variation of the mixed traffic (Tesoriere, Campisi, Canale, & Zgrablić, 2018). The sensitivity of the results was tested by changing the values of the input variables. We usually talk about analysis for (future) scenarios, where a scenario represents one of the possible combinations of values. The sensitivity analysis, therefore, provides useful indications regarding the riskiness of a project or the definition of a scenario and the sources from which it derives. In this research, it was considered appropriate to maintain the traffic composition as a constant variability and vary its composition in terms of pedestrian flow. The observed pedestrian flow was 1200 ped/h. This value was used for the 1st Scenario. The other two scenarios were 100% and 200% raised from the first scenario respectively. For all three scenarios the cyclist flow was 450 cycle/h and the vehicular flow was 400 veh/h. As far the composition of the pedestrians for all scenarios there were: young male 6%, adult male 30%, elderly male 13%, young female 7%, adult female 28%, elderly female 13%, wheelchairs 2% and mums with child 2%. The composition of vehicular traffic flow was: light vehicle 55%, heavy vehicle 20%, buses 20% and motorcycle 5%. The scenarios take into consideration an increase in the percentage of the elderly, as the figures shows for Greece.

4 DESCRIPTION

The monitored area is located in Kalamaria, a Municipality of the Thessaloniki Greater Area (TGA), located about 7 kilometres southeast of downtown Thessaloniki. It is the second largest Municipality of the TGA (actually about 91600 habitants) as well as one of the largest in Greece. It is characterized by the presence of two pedestrian infrastructures called Komninon and Metamorfoseos respectively. Counts were carried out during April 2019, during typical weekdays (not weekends), morning hours and under fine weather conditions. The counts were related to the traffic flows, their composition and their speed.

Metamorfoseos street is characterized by an infrastructure with lanes dedicated to the bicycle flow (red street) that intersect with the lanes dedicated to pedestrians (brown area). Figure 1 presents the Metamorfoseos pedestrian street. The analysis of the LOS was carried out by subdividing the analysed street into several numbered sub-areas. The area is also characterized by the presence of numerous commercial activities. In proximity of the under study area a metro station is being built which will allow an increase in the pedestrian flow circulating in the area when the metro system will be operational.

The evaluation in terms of the LOS value based on the flow composition allows the local authority to be able to support some choices relating to the area such as the location of areas for pedestrians parking or the mitigation of potential conflict points at the road crossings .The research shows how, as pedestrian flow varies, some areas of the infrastructure are characterised by reduced road reduce their LOS for the same traffic composition. This is due to the greater iteration of pedestrians as a result of the geometry of the area and the potential conflicts between the different traffic components.

5 RESULT OF THE RESEACH

The evaluation of the LOS was introduced for the first time in motorized traffic in the 60s in the Highway Capacity Manual (HCM, 2010). The calculation of the LOS is based on the assumption that drivers perceive that the low traffic volumes have a higher quality and therefore a higher LOS.

To determine the LOS, the operating speed and the volume/capacity ratio were used. Based on these criteria, six levels were distinguished, from A to F. The analysis of pedestrian areas is made in such a way to include the evaluation of the PLOS. This assessment is done through methodologies that influence the planning, design and operational aspects of transport projects, as well as the allocation of limited financial resources in competing transport projects.

Figure 1.   Metamorfoseos pedestrian street.

Table. 1. Comparison of different scenarios considering LOS values.

LOS	comments	Avg Speed (m/s)	Density (ped/m2)	
A	Free flow No conflicts	>1.3	<0.27	1st Scenario
B	Normal walking speed Minor conflicts	1.27	0.43-0.31	
C	Restricted flow Some conflicts	1.22	0.72-0.43	2nd Scenario
D	Conflict Walking speed restricted	1.14	1,08-0.72	
E	More Conflict Walking speed restricted	0.76	2.17-1.08	
F	Shuffing and buching Extreme restriction of speed	<0.76	>2.17	3rd scenario

The evaluation of the PLOS requires the acquisition of geometric-functional data character-izing the infrastructure, such as the length, width, pedestrian density and flow rate. Table 1 present the comparison of different scenarios considering LOS values. The eastern part of the infrastructures is the one where

LOS varies. It must be mentioned at this point that the special area is characterised by high density of interactions between the dedicated lanes. A weak variation in the service level is also found in the initial part west of the infrastructure of the cycle-pedestrian lanes.

6 CONCLUSIONS

The present research highlights the variation in the pedestrian Level of Service (LOS) through the evaluation of a sensitivity analysis as to the variation of the pedestrian flows, maintaining constant the composition of the traffic and the cycle and motorized flows.

This analysis lays the foundations for a more accurate assessment of the area under study wherein the coming years a metro station will be built. The results of the analysis can help towards the design of a sustainable city mobility system in case of increased pedestrian flows. The findings of the paper are presented in Table II (3 scenarios) taking into account the prob-lem of congestion as well as the conflict points between motorized and non-motorized traffic.

REFERENCES

Fruin, J. J. (1971). Pedestrian planning and design.
HCM. (2010). Manual. Transportation Research Board of the National Academies.
Helbing, D., & Molnar, P. (1995). Social force model for pedestrian dynamics. Physical review E, 51 (5).
Nikiforiadis, A., & Basbas, S. (2019). Can pedestrians and cyclists share the same space? The case of a city with low cycling levels and experience. Sustainable Cities and Society, 101453.
Panagopoulos, T., Tampakis, S., Karanikola, P., Karipidou-Kanari, A., & Kantartzis, A. (2018). The Usage and Perception of Pedestrian and Cycling Streets on Residents' Well-being in Kalamaria. Greece. Land, 7 (3), 100.

PTV Group. (2015). PTV Vissim. Retrieved from PTV Group. http://vision-traffic.ptvgroup. com/enus/
products/ptv-vissim/use-cases/junction-geometry.

PTV Group. (2011). PTV Viswalk. Karlsruhe: PTV Group.

Tesoriere, G., Campisi, T., Canale, A., & Zgrablić, T. (2018). The Surrogate Safety Appraisal of the
Unconventional Elliptical and Turbo Roundabouts. Journal of Advanced Transportation.

Pedestrians, Urban Spaces and Health – Tira, Pezzagno & Richiedei (eds)
© 2020 Taylor & Francis Group, London, ISBN 978-0-367-46171-3

Urban spaces and mobility in Makkah city: Ordinary organization and big events

R. Bahshwan, R. De Lotto & C. Berizzi
Department of Civil Engineering and Architecture, University of Pavia, Italy

ABSTRACT: The paper aims to introduce research that is under investigation by an international and multidisciplinary group at the University of Pavia. The primary purposes are specifically addressed to the city of Makkah in the Kingdom of Saudi Arabia. Key objectives of the research are related to the transportation city plan, sustainable mobility organization, pedestrian paths and spaces, urban spaces design, and crowd management. What makes the case of Makkah interesting is the coexistence of different characteristics of the city and different typologies of people who live in, for long or short periods, according to the big events that occur yearly. Therefore, starting from the necessity to have a coordination between the ordinary city life and the high density of occasional but assured city users, the research seeks to recognize the opportunities of urban design that the fluxes and crowd distribution optimization can propose. As the first step of a larger investigation, the authors will focus on the structure of the research and the first results.

1   INTRODUCTION

Authors convened an interesting field of multidisciplinary research starting from the description of a unique case study that is represented by the City of Makkah. One of the authors come from this city, and he was able to highlight the merging of transportation planning issues together with urban planning and urban design issues in solving some evident problems Makkah is facing. On the one hand, the city could be analyzed as a "permanent big event city" considering that each year at least two significant events take place (Flyvbjerg, 2014). On the other hand, due to the huge number of visitors that Makkah receives for a short period, there are many parts of the city that are used only for a little time (Alamoudy, 2017). This fact represents an interesting field of research in order to define strategies and design proposals to make Makkah take advantage of the events for improving its mobility system (with particular reference to pedestrian and citizens) and the quality of its urban spaces. The so called "big events" consider at least: Expo and Olympic Games. In Shanghai Expo 2010 there were 73 million visitors, in Milan Expo 2010 they were 22, in Dubai Expo 2020 the forecast is 25 million. In 2012 London Olympic games the visitors were 9 million, in Rio 2016 they were 6 million. In Makkah the 2030 forecast is about 37 million visitors every year.

2   OBJECTIVES OF THE PAPER

In the paper, authors synthesize a research that aims to propose solutions to a set of problems that nowadays represent serious issues in Makkah city; one basic passage is to systematize these aspects into disciplinar categories such as: society groups definition, traffic management, urban planning, urban design. This is useful to deeply investigate the different facets and the causal relations of the phenomena, to propose possible planning and operative instruments, to connect strategies and actions to the legislative structure of Makkah. From traffic plans, to traffic management, to sustainable mobility plans, to specific urban design actions addressed to different kind of city users. To frame the problem, the first step is to analyze the context in order to understand which similarities and differences occur in Makkah situation in comparison with other big events cities (Biancardi, De Lotto, and Ferrara, 2008). The context is described from the geographical urban development point of view and the city users categories.

3 RESEARCH APPROACH

The wider research the paper is about is based on the original features of Makkah big events, the relation between citizens and temporary city users, the opportunities that certain strategical urban structure decisions may introduce, make the approach necessarily multidisciplinary. For what has been previously cited, many key issues must be faced to analyze the context and to (try to) insert them into general worldwide well-known approaches or at least categories. The methodology of the whole research follows these steps: 1) description of the city of Makkah development; 2) description and scale definition of the events that every take place in Makkah; 3) definition of the different "city users" that live in the city (for short or long periods); 4) qualitative definition of their mobility preferences; 5) definition of the main paths/ directions; 6) definition of the main multimodal transportation means (for long and short distances); 7) definition of the main civic/technical instruments pertinent to the local administrative law to reach the goals; 8) similar case study guideline; 9) comments and conclusions. In this paper, the authors develop points 1), 2), 3) and 8).

4 DESCRIPTION

4.1 *The city of Makkah (Mecca)*

Makkah is the center of the Islamic faith. It is the birthplace of Islam and Prophet Mohammed, a direction for all Muslims in their daily prayers, a destination to accomplish the fifth Islam pillar (Al-Hajj or pilgrimage), and possession the Holy Sites (the Holy Mosque, Mina, Arafat, Muzdalifah) (Amirahmadi and El-Shakhs, 1993). The city is located on the west side of the Kingdom of Saudi Arabia, elevated between 250 to 350 meters above sea level, and surrounded by historical and huge mountains. During summer, the temperature in Makkah reaches a scorching 45+ degree Celsius while in the winter months, the temperature hovers at comfortable 24 degree Celsius (Habeebullah, 2012). The past development plans for Makkah have played a role in shaping the city as it is today. In 1512, the city area was about 0.59 km^2. (Amirahmadi and El-Shakhs, 1993). After establishing the Kingdom of Saudi Arabia, particularly during the discovery of oil in 1940c, the city dramatically has transformed (Figure 1). The last fifteen years have been witnessing phenomenal investment in developing and improving the Holy Sites. Many historical residential areas demolished and mountains were flattened in order to expand the Holy Mosque as well as constructing massive hotels, restaurants, sanitary facilities, and infrastructures (Mekki, 1988). Now the city area has expanded approximately 1116 km^2 according to the Ministry of Municipal and Rural Affairs reports.

4.2 *Big events in Makkah: Main destinations*

Millions of Muslims from more than 183 countries are attracted to Makkah either to visit the Holy Mosque or to perform a pilgrimage (Rahman, 2017). There two types of pilgrimage (Al-Hajj and Umrah), and each one has a specific itinerary. Al-Hajj, the largest gathering occurs once every year over six days during the last month of the Islamic calendar (Dhu Al-Hijjah). It is an obligation for all capable (physically and financially) Muslims at least once in their lifetime. This ritual takes place in all the Holy Sites. Umrah (lesser pilgrimage) is optional and can be performed anytime during the year (the peak period in Ramadan), and is restricted to visiting the Holy Mosque. During these periods of Al-Hajj and Umrah, the primary concerns

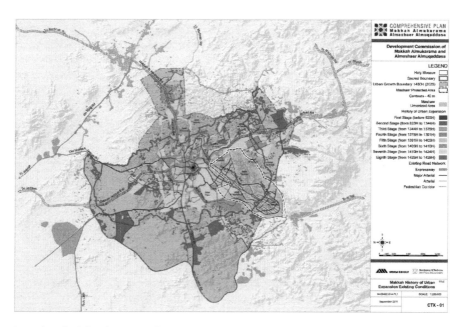

Figure 1.   Historical development of Makkah.

to the Saudi government relate to the personal safety of the pilgrims. In the past pilgrims have been the victims of pedestrian stampedes, random violence, terrorist attacks, natural disasters (e.g., flood), and accidents (e.g., fires). One of the leading causes of these disasters is over-crowding. However, Umrah's pilgrims set up to grow from eight million to thirty million by 2030. The unusual rapid increasing number of Makkah's visitors and pilgrims occurs due in part to the availability of air travel. During Al-Hajj, for the first time, in 1950c, the pilgrims' number was exceeded a million. Then again, in 2012, according to Al-Hajj Statistics, it became around 3,161,573 pilgrims.

This number is expected to grow more than four million in the next twenty years. Not only Makkah stresses growth with the number of visitors and pilgrims but also the inhabitants. In 1843, the population was about 19,000 people. Currently in 2015, Emirate of Makkah Province reported that it reached to 1,578,722 people (Alamoudy, 2017; Amirahmadi and El-Shakhs, 1993).

4.3   City users' main mobility in Makkah

Mohammed Badabaan (2013) points out that most of the pilgrims prefer using buses rather than walking. However, Amen Younis (1982) states that some pilgrims aim to walk from one Holy Site to another in order to follow Prophet Mohammed, and the others have various reasons such as avoiding traffic jam and high tickets prices. Furthermore, there is a minimum allocation for streets and public spaces results in a conflict between vehicles and pedestrian. The pedestrian paths capacity cannot hold the influx of a high number of pilgrims. The study of pilgrims scheduling approach, which prescribes specific routes and times for pilgrims to flow, purposes to solve the pedestrian congestion during Al-Hajj. The city users' scheme is principally composed by: 1) residents, 2) visitors, 3) pilgrims for Al-Hajj, 4) pilgrims for Umrah. Residents in recent times have been moved from the city center to new suburban settlements and new towns. In these new settlements also workplaces were moved, so for residents, there is a preference for private car mobility. Visitors come to Makkah for different reasons but mainly for tourism (cultural or religious). They stay mainly in the hotel districts (in the city center) or by parents (all around the city). Pilgrims for Al-Hajj arrive in Makkah a little before the "6 days" in the last months Dhu Al-Hijjah; before the "6 days," they mainly stay between the second and third ring road districts. They mainly move by public transportation and taxi. Pilgrims for Umrah come to Makkah during the whole year, and the peak period is in Ramadan. They mainly stay close to the Holy Mosque (in the city center), and they can move with private cars or public transportation.

5 CASE STUDY REFERENCES

In big events, pedestrian mobility spaces have a dual function. During the event, they are used for a large number, and in the rest of the year for residents. This mobility infrastructure plans to transit large flows quickly and safely; however, the quality of space becomes secondary importance. In particular, the Olympics 2016 in Rio de Janeiro, Brazil, and 2012 in London, the spaces designed for visitors as technical elements without any architectural value. Once the event was over, they were removed or modified. Unlike the case of the Universal Forum of Cultures in Barcelona that took place in 2004 and Milan Expo in 2015, the spaces created for the event, and set a goal for the urban transformation process. These former cases express a trend that takes place in the most important metropolitan areas where open spaces are destined to pedestrianization and becoming strategic for urban quality considering even the environmental, economic and social dimensions (Gianfrate and Longo, 2017; Clemente, 2017; Berizzi, 2018).

6 CONCLUSIONS

The first step of the research aims to recognize some priority elements in order to address Makkah mobility planning issues and Makkah urban design attention considering that a huge number of different kind of city users arrive in Makkah and they must coexist with citizens and their needs. So the mobility urgency can be exploited to reach a better urban quality and to make efficient mobility systems available all the year. The first step of the research focused on understanding the main urban structure according to its flexible density (De Lotto and Morelli di Popolo, 2015) and destination significance. The qualitative analysis will be further deepened with quantitative data that can be acquired with cooperation with the Makkah Municipality. Then, it will be possible to define with appropriate details the main axes that are prior in pedestrian mobility.

REFERENCES

Alamoudy, S., (2017). Urban Transformation Through Creativity: Applying the Creative City Concept to Makkah. 250–288.
Amirahmadi, H., & El-Shakhs, S., (1993). Urban development in the Muslim. 37–58.
Badabaan, M., (2013). Estimation of trips generated from pilgrims' hotels and homes in Makkah during the seasons of Ramadan and Hajj 1434. 16–41.
Berizzi, C., Piazze e Spazi Collettivi, Il Poligrafo, Padova 2018 – in Italian.
Biancardi, A., De Lotto R., Ferrara, A., (2008) Urban services localization and optimal traffic distribution: Users oriented system, Journal of Urban Planning and Development, American Society of Civil Engineers, Volume 134, Issue 1.
Clemente, M., Re-Design dello spazio pubblico, Franco Angeli Editore, Milano 2017 – in Italian.
De Lotto, R., Morelli di Popolo, C. (2015). Complex, adaptive and hetero-organized urban development: the paradigm of flexible city. pp.22–26. In The 6th International Multi-Conference on Society and Information Technology.
Flyvbjerg, B. (2014). What you should know about megaprojects and why: An overview. Project Management Journal, 45, 6–19.
Gianfrate, V., & Longo, D., (2017). Urban micro-design. Franco Angeli – in Italian.
Habeebullah, Turki, (2012). Application of the use of catalytic converter to reduce air pollutants inside Makkah tunnels. 24–31.
Mekki, Z. A., (1988). Transportation Problems in The City of Makkah Outside the Period of Hajj. 1–2.
Rahman, J., (2017). Mass Gatherings and Public Health: Case Studies from the Hajj to Mecca. 386–387.
Younis, Amen, (1982). Pedestrian Paths in Al-Mashaer Al-Muqaddasa 1402, Sami Angawi. 8.

Pedestrians, Urban Spaces and Health – Tira, Pezzagno & Richiedei (eds)
© 2020 Taylor & Francis Group, London, ISBN 978-0-367-46171-3

Increasing urban walkability: Evidences from a participatory process based on spatial configuration analysis

P. Pontrandolfi
Department of European and Mediterranean Cultures, University of Basilicata, Matera, Italy

B. Murgante, F. Scorza, R. Carbone & L. Saganeiti
School of Engineering, University of Basilicata, Potenza, Italy

ABSTRACT: The research analyses the theme of walkability in the western part of Potenza municipality. It combines spatial configuration analysis and evidence-based partecipatory process developed in the project "Active citizenship for Sustainable Development of the Territory" (CAST). The paper compare results of the public participation process concerning walkability and neighborhood acceesibility with the results of the configurational analysis based on space syntax. During this research expereince, a cognitive framework has been defined both adopting traditional approaches, and, in order to increase the participation, using new information technologies and social networks. The data that emerged were revised and evaluated for the definition of feasibe strategies for the improvement of walkability, accessibility to the services and infrastractures and, more generally, the neighborhood liveability.

1   INTRODUCTION

Assuming the concept of walkability, as the possibility for citizens to move on foot is essential to provide them a modal choice in moving. In other words: choose whether to move on foot or not. Walking should be considered as a valid alternative to motor vehicles. In today's sense, however, walking is seen as a limitation and this is demonstrated by the fact that the possibility of moving on foot is often reserved for a few categories of users or, in the worst cases, a few areas.

The concept of walkability (Cervero and Duncan 2003, Ewing and Handy 2009, Carr *et al.* 2010, Blečić *et al.* 2015, Blečić, Canu, *et al.* 2016, Blečić *et al.* 2017) goes beyond accessibility to urban services, but implies a discussion on the quality of spaces and the ability to invest in pedestrian mobility in urban areas. Walkability should be understood as one of the main factors of urban capacity (Nussbaum *et al.* n.d., McCann 2002).

The quality of urban life does not represent the quality of life of individuals living in a certain geographical area, but the quality of the urban environment that influences development and the possibilities of choice and action, according to individual needs and wishes (Arras *et al.* 2014, Blečić, Cecchini, *et al.* 2016).

In this case, a participatory process was used (supported in the final phase by an analytical analysis (Ssx)) in which citizens were able to express strengths and weaknesses, opportunities and needs of their neighborhood in order to improve the level of viability of the neighborhood and thus improve the quality of life.

2   OBJECTIVE OF THE PAPER AND STUDY CASE

The paper's object is the application of the configurationally analysis theories or Space syntax analysis (Ssx) to the study case of Potenza municipality western zone, with the aim of assessing the pedestrian accessibility of the area.

As expressed in the theories of Bill Hillier (Hillier *et al.* 1976, Hillier and Tzortzi 2006), founder of Ssx, the movement within public space consists of a portion of movement attracted, influenced by the presence of activities and services, and a portion determined exclusively by the spatial configuration, the natural movement.

With reference to these theories, this work aims to study natural movement in a neighborhood by investigating how spatial configuration can affect pedestrian movement and accessibility within the neighborhood. The results of the SSX obtained on the state of fact with those obtained as a result of the insertion of new project arcs or the improvement of existing ones will be analyzed.

The study area includes the western part of the Potenza municipality, that is a mountain city situated in Basilicata region (southern Italy). Over the years, the city has welcomed several participatory activities, in most cases with a bottom-up approach (Lorusso *et al.* 2014, Amato *et al.* 2015, Rocha *et al.* 2015, 2016, Sassano *et al.* 2017), and in other cases promoted directly by local authorities (Murgante *et al.* 2011, Scorza and Pontrandolfi 2015).

The area investigates (Fig.), includes the districts of *Poggio Tre Galli*, the *G area* and the *"study center"* area.

Figure 1.   Location of the study area: the neighbourhood of *Poggio Tre Galli*, the *G area* and the district called "Study-Centre".

Poggio Tre Galli and *G area* are predominantly residential neighborhoods with a high density building, while in the study center there are many higher education institutions and lower grade schools and many unpublished areas.

3   SPACE SYNTAX ANALYSIS

The SSX is characterized by the central role given to urban space in the face of settlement phenomena. It uses quantitative models to measure the effects of the spatial configuration on the physical properties of an urban aggregate and on its immaterial variables and recognizes in space the essential reasons for the phenomena that take place on it, so as to be able to interpret and understand them, but above all to be able to predict and simulate them. This approach allows to interpret and understand the internal geography of an urban aggregate; to suggest uses and destinations of the land congruent with the potential offered by the articulation of the urban space; to simulate the effects of the transformations in the project on the variables, material and immaterial, of the system, becoming a decision support system (DSS). The Space Syntax analysis is based on three assumptions: (i) the structure of the urban space influences the phenomena that take place within it; (ii) the perceptual appreciation of the space influences the behavioural choices within it; (iii) the grid of urban paths is considered as the constitutive structure of the urban space. Among the different methodologies for the configurational analysis in this work, the axial analysis has been used. The Ssx was carried out on the road network of the vehicular and pedestrian area using the DepthMapXnet software.

The connectivity (Knight and Marshall 2015), integration and choice indices (Li *et al.* 2017) have been calculated and compared using the statistical indices of mean, standard deviation, variance; minimum and maximum values, Gini coefficient.

4 RESULTS

Following numerous participatory meetings of the CAST project, it was possible to define the cognitive framework of the study area. Among the various themes analyzed, attention was focused on that of the liveability of the neighborhood and therefore also of the walkability. From the analysis of the problem tree, built with the citizens also with walks in the areas of interest, several problems emerged that denote a discontinuous pedestrian network and lacking both quantitatively and qualitatively. In particular, the presence of architectural barriers, discontinuity of sidewalks, restricted sections of routes, lack of safety measures, dangerous pedestrian crossings, poor maintenance and cleaning. All critical issues that negatively affect the accessibility of the area. On the basis of these critical points identified, the first Ssx analysis was carried out then, once the project interventions had been defined, the analysis was re-performed. The project interventions mainly concerned the improvement of existing arches with the enlargement of the section or the construction of new road arches (for a detailed description see (Carbone *et al.* 2018)). The results of the two analyses are summarised in Table 1 and discussed in the next paragraph.

Table 1. Statistical values obtained for the three indices calculated with the Ssx. PRE indicates the situation before the project and POST the situation after the project.

STATISTIC	Connettivity index		Integration index		Choice index	
	PRE	POST	PRE	POST	PRE	POST
Number of features	236	236	236	277	236	277
Mean	5.84	5.84	1.95	1.99	0.05	0.04
Standard deviation	3.96	3.96	0.67	0.61	0.08	0.07
Variance	15.69	15.69	0.45	0.38	0.01	0.00
Minimum	1.0	1.0	0.33	0.33	0.0	0.0
Maximum	21.0	21.0	3.54	3.57	0.39	0.41
Coeff. Gini	0.37	0.37	0.19	0.17	0.66	0.67

5 DISCUSSIONS AND CONCLUSIONS

From the statistical analysis it is possible to quickly compare the PRE and POST situations: the connectivity index does not undergo changes with the addition of new pedestrian paths and is similar in both hypotheses; as regards the integration index increases the maximum value that means greater accessibility of the area (shorter distance of travel between the various arches) as evidence of this decreases the coefficient of gini (thus decreases the inhomogeneity of the area); in the index of choice slightly increases the maximum value after the insertion of new arcs as evidence of a greater alternative of routes.

In conclusion, the Ssx has been used to validate and deepen the results of a participatory process and has therefore allowed to explain in a scientific way when it emerged from the analysis carried out during the laboratory. It also made it possible to understand how to operate on the different parts that make up the study area and to evaluate the effects of the project hypothesis defined with the participatory activity.

To understand the spatial dynamics and have a complete framework of pedestrian mobility, it is necessary to integrate the more classical methodologies that are usually used and that allow to analyze the attracted movement and the equipment and quality of the structures for pedestrian movements with analysis on the spatial configuration.

REFERENCES

Amato, F., Bellarosa, S., Biscaglia, G., Catalano, L., Graziadei, A., Metta, A., Murgante, B., Olivetti, M.L., Passannante, P., Percoco, A., Sassano, G., and Scaringi, F., 2015. "Serpentone Reload" an Experience of Citizens Involvement in Regeneration of Peripheral Urban Spaces. Springer, Cham, 698–713.

Arras, F., Cecchini, A., Ghisu, E., Idini, P., and Talu, V., 2014. Perché e come promuovere la camminabilitá urbana a partire dalle esigenze degli abitanti piú svantaggiati: il progetto "Extrapedestri. Lasciati conquistare dalla mobilità aliena!" 9° Congresso Città e Territorio Virtuale, Roma, 2, 3 e 4 ottobre 2013, 185–196.

Blečić, I., Canu, D., Cecchini, A., Congiu, T., and Fancello, G., 2016. Factors of Perceived Walkability: A Pilot Empirical Study. Springer, Cham, 125–137.

Blečić, I., Canu, D., Cecchini, A., Congiu, T., and Fancello, G., 2017. Walkability and Street Intersections in Rural-Urban Fringes: A Decision Aiding Evaluation Procedure. Sustainability, 9 (6), 883.

Blečić, I., Cecchini, A., Congiu, T., Fancello, G., and Trunfio, G.A., 2015. Evaluating walkability: a capability-wise planning and design support system. International Journal of Geographical Information Science, 29 (8), 1350–1374.

Blečić, I., Cecchini, A., Fancello, G., Talu, V., and Trunfio, Giuseppe, A., 2016. Camminabilità e capacità urbane: valutazione e supporto alla decisione e alla pianificazione urbanistica. TERRITORIO ITALIA.

Carbone, R., Saganeiti, L., Scorza, F., and Murgante, B., 2018. Increasing the Walkability Level Through a Participation Process. Springer, Cham, 113–124.

Carr, L.J., Dunsiger, S.I., and Marcus, B.H., 2010. Walk Score^{TM} As a Global Estimate of Neighborhood Walkability. American Journal of Preventive Medicine, 39 (5), 460–463.

Cervero, R. and Duncan, M., 2003. Walking, bicycling, and urban landscapes: evidence from the San Francisco Bay Area. American journal of public health, 93 (9), 1478–83.

Ewing, R. and Handy, S., 2009. Measuring the Unmeasurable: Urban Design Qualities Related to Walkability. Journal of Urban Design, 14 (1), 65–84.

Hillier, B., Leaman, A., Stansall, P., and Bedford, M., 1976. Space syntax. Environment and Planning B: Planning and Design, 3 (2), 147–185.

Hillier, B. and Tzortzi, K., 2006. Space Syntax: The Language of Museum Space. In: S. Macdonald, ed. A Companion to Museum Studies. Blackwell Science Ltd, 282–301.

Knight, P.L. and Marshall, W.E., 2015. The metrics of street network connectivity: their inconsistencies. Journal of Urbanism: International Research on Placemaking and Urban Sustainability, 8 (3), 241–259.

Li, X., Lv, Z., Zheng, Z., Zhong, C., Hijazi, I.H., and Cheng, S., 2017. Assessment of lively street network based on geographic information system and space syntax. Multimedia Tools and Applications, 76 (17), 17801–17819.

Lorusso, S., Scioscia, M., Sassano, G., Graziadei, A., Passannante, P., Bellarosa, S., Scaringi, F., and Murgante, B., 2014. Involving Citizens in Public Space Regeneration: The Experience of "Garden in Motion". Springer, Cham, 723–737.

McCann, E.J., 2002. Space, citizenship, and the right to the city: A brief overview. GeoJournal, 58 (2/3), 77–79.

Murgante, B., Tilio, L., Lanza, V., and Scorza, F., 2011. Using participative GIS and e-tools for involving citizens of Marmo Platano–Melandro area in European programming activities. Journal of Balkan and Near Eastern Studies, 13 (1), 97–115.

Nussbaum, M.C., Sen, A., and World Institute for Development Economics Research., n d. The quality of life.

Rocha, M.C.F., Pereira, G.C., Loiola, E., and Murgante, B., 2016. Conversation About the City: Urban Commons and Connected Citizenship. Springer, Cham, 608–623.

Rocha, M.C.F., Pereira, G.C., and Murgante, B., 2015. City Visions: Concepts, Conflicts and Participation Analysed from Digital Network Interactions. Springer, Cham, 714–730.

Sassano, G., Graziadei, A., Amato, F., and Murgante, B., 2017. Involving Citizens in the Reuse and Regeneration of Urban Peripheral Spaces. Springer, Cham, 193–206.

Scorza, F. and Pontrandolfi, P., 2015. Citizen Participation and Technologies: The C.A.S.T. Architecture. Springer, Cham, 747–755.

Pedestrians, Urban Spaces and Health – Tira, Pezzagno & Richiedei (eds)
© 2020 Taylor & Francis Group, London, ISBN 978-0-367-46171-3

Auditing streets' pedestrian compatibility: A study of school sites' requalification

M. Ignaccolo, G. Inturri, G. Calabrò, V. Torrisi, N. Giuffrida & M. Le Pira
University of Catania, Catania, Italy

ABSTRACT: This paper aims at promoting sustainable mobility in urban areas, by focusing on the assessment of pedestrian routes compatibility in proximity of schools and the identification of both short and long term interventions. The case study is the city of Acireale, a small town of 50,000 inhabitants in Sicily (Italy), characterized by the presence of school facilities scattered throughout the whole urban area. Based on previous research, a walkability index was calculated and thematic maps were elaborated in a GIS environment. A case study of educational centre with a low pedestrian compatibility index has been analysed in detail and different measures of intervention have been proposed, aiming at ensuring continuity, safety and pleasantness of pedestrian paths. The overall aim is to provide decision-makers with a planning-support method that can aid them to decide the priority of investments, based on street centrality and pedestrian compatibility.

1   INTRODUCTION

A contribution to sustainable city in economic, social and environmental aspects is identified in walking. Moreover, many studies have reported the health benefits of walking, as a practical solution to the alarming epidemic of obesity and chronic diseases (Doorley et al., 2015; Mueller et al., 2015). By the way, active school transport has the potential to contribute substantially to physical activity and health (Faulkner et al., 2009). Walking or cycling to school has been schown to be associated with a number of variables at the invidual, cultural, social level and physical environment factors (Chillon et al., 2011; van Loon and Frank, 2011).In the south of Italy, cities are characterized by high motorization rates and car-dependence even for short distance journeys. This often occurs since pedestrian facilities are not equipped to guarantee the accessibility of the areas of interest, such as public buildings, workplaces, educational and sports facilities, commercial areas and public transport nodes. Recent literature has considered various indicators of walkability (e.g., connectivity, safety, comfortability, accessibility, and convenience), which yield numeric scores for auditing street walkability (Kurka et al., 2016; Moura et al., 2017). It has been clear that street walkability should be better described using a composite set of indices rather than single one (Maghelal and Capp, 2011). Several factors should be taken into consideration to evaluate the walkability of a pedestrian infrastructure, concerning pedestrians' flows, density of activities, the degree of roads connection, the diversification of land use and the presence and maintenance of pedestrian infrastructures. In this regard, schools are sites where pedestrian facilities should be designed to be used by all individuals, especially by the "weaker" categories, such as children. They are characterized by the presence of significant pedestrian flows concentrated at particular times of the day and interventions must pursue the objective of protecting students' flows through the construction of an interconnected network of sidewalks and pedestrian crossings, the removal of architectural barriers and the reduction of vehicular speeds. However, in order to establish priority of interventions, it is important to have detailed spatial data regarding the characteristics of pedestrian environment (e.g. sidewalk provision, access ramps, gradient, lighting) and to define appropriate indicators of walkability.

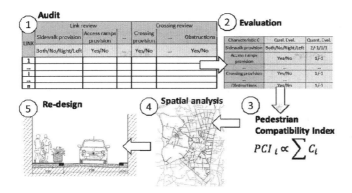

Figure 1. Research framework.

The paper proposes a study aimed at promoting sustainable mobility in urban areas, by focusing on the assessment of pedestrian routes compatibility in proximity of schools and the identification of both short and long term interventions. A GIS-based tool has been used for spatial analysis based on data collected during field surveys. This allowed to create a thematic map related to pedestrian compatibility for educational facilities' zones, and consequently, to identify requalification proposals.

2 RESEARCH APPROACH

The research approach used in this study relies on an on-field audit aimed at gathering data regarding pedestrian environments, and a spatial analysis of walkability conditions. In this respect, a walking audit tool was built based on the Pedestrian Environment Review System (PERS) developed by TRL (Transport Research Laboratory) (Allen and Clark, 2007). In particular, different characteristics of road links composing a pedestrian network, i.e. sidewalk provision and width, access ramps, gradient, obstructions, lighting, tactile information, surface quality, continuity and parking provision were assessed, as well as pedestrian crossings characteristics, like ramps, maintenance, excessive lenght and obstructions. These characteristics were qualitatively rated (e.g. low/high gradient), and then converted to quantitative positive/negative values to allow comparability among them (e.g. -1 for high gradient, +1 for low gradient). Then, a Pedestrian Compatibility Index encompassing the different characteristics was calculated. Based on this analysis, a spatial database was created and thematic maps elaborated in a GIS environment. From the spatial analysis, based on street centrality and pedestrian compatibility, some critical areas needing improvements emerged and design solutions proposed. The overall research framework is summarized in Figure 1.

3 DESCRIPTION: CASE STUDY AND AUDIT RESULTS

The case study is Acireale, a small touristic town of 50,000 inhabitants in Sicily (Italy). Main problems related to mobility in Acireale include lack of accessibility and efficiency of public transport, insufficient infrastructures dedicated to cycling and walking, high level of road congestion, also due to infrastructural inefficiencies of the road network, with consequent air pollution and low levels of livability. Besides, Acireale is also characterized by the presence of school institutions scattered throughout the urban area, both in peripheral zones and near the historical center. In this respect, several cases of educational centres with a low pedestrian compatibility index have been analysed more in detail and different measures of intervention proposed to improve the quality of pedestrian paths. The design solutions aim

Figure 2.   Average PCI for educational facilities' zones in the urban area.

at ensuring continuity, safety and pleasantness of pedestrian paths, by coordinating multiple interventions of sidewalk enhancements, removal of barriers represented by on-street parking and traffic calming measures such as raised pedestrian crossing. Results of the audit conducted show that the areas on which it is necessary to focus most the attention are those located in the centre-south of the urban area: among these areas, those located in the historical center are characterized by typical narrow streets, and show as principal issue the absence of the sidewalk. Anyway they can be considered easily predisposed for a pedestrianization since they are mainly reserved for residents and they are provided with speed limit signs. Among the other streets with a low PCI, definitely more peripheral and equipped with large enough road sections to think of a pedestrian requalification intervention that does not affect the current viability, some are caracterized by poles attractors of mobility of unprotected users and in particular by the presence of important educational facilities. Based on the data on Points of Interest of the Urban Mobility Plan, a spatial analysis was conducted to identify the study area for each school to focus attention on the redevelopment of those with a bad pedestrian compatibility. For this purpose, a 300 meter buffer was drawn around each educational facilities and the average value of the PCI of the access roads was assigned to it (see Figure 2).

4   REQUALIFICATION PROPOSAL

In this section we present the significant case study of the pedestrian paths' requalification around a large schools complex located on the edge of the Acireale's urban fabric. A survey of the site is reported in Table 1 and Figure 3.

Table 1.   Critical issues detected.

Element	Rating	Critical issues
Sidewalks	Poor	No sidewalk at the entrance to the school (**Figure 3a,b**); continuity of paths not ensured. Surface's quality in general acceptable
Gradient	Good	None
Curb ramps	Poor	No crossing equipped with curb ramps
Pedestrian crossings	Fair	Pedestrian crossings present: length in **Figure 3c** acceptable but interrupted by a traffic island, while in **Figure 3a** excessively long
Mobility barriers	Fair	Some obstacles to pedestrian mobility (**Figure 3b**) narrowing the pedestrian path on a sidewalk of width inadequate in itself
Parked cars	Poor	Although parking is not allowed in front of the
Lighting	Good	None
Transit stops	Fair	Bus stop near the school present, but not indicated by proper signs

Figure 3.   Current state and main critical issues identified.

Figure 4.   Layout of the area subjected to requalification measures.

The proposed interventions are shown in Figure 4 and described as follows:

1. creation of raised pedestrian crossings as a traffic calming measure aimed at reducing vehicle speeds near the entrances to schools;
2. two-color marking of pedestrian crossings and construction of curb ramps to connect the sidewalk with the road pavement;
3. curb extension at the intersection with insertion of parking bollards;
4. construction of the sidewalk in front of the entrance to the school;
5. broadening of sidewalks in the entire intersection area to guarantee, where possible, a useful pedestrian path with a width of 2.00 m;
6. reduction of the roundabout's entry radius;
7. creation of a lifebuoy island (width of 2.00 m) along the crossing, to allow a comfortable refuge for pedestrians.

5   CONCLUSIONS

Urban infrastructures should be designed to be used by all individuals, especially by the "weakest" categories. The removal of architectural barriers is an act of civilization which must be addressed especially in proximity of schools and educational centres, characterized by significant pedestrian flows concentrated in particular times of the day. Interventions should pursue the key objectives of ensuring continuity, safety and pleasantness of pedestrian paths around school complexes through coordinated interventions of sidewalk enhancements, traffic calming measures, removal of barriers represented by on-street parking, etc. A greater presence of pedestrian crossings and suitable curb ramps improves access between the sidewalk and roadway to individuals who have mobility restrictions and ensures the continuity of the pedestrian itinerary, while an adequate width of the sidewalk allows a safe and comfortable passage even of people in wheelchairs. Thanks to the greater spaces dedicated to pedestrians and the illegal parking prevention as well as the higher safety of raised pedestrian crossings, made shorter and more visible by curb extensions, it is possible to pursue the recovery of the liveability of urban environment.

REFERENCES

Allen, D., Clark, S., 2007. New directions in street auditing: lessons from the PERS audits. In Int. Conf. on Walking and Liveable Communities, 8th, 2007, Toronto, Ontario, Canada.

Chillon, P., Evenson, K., Vaughn, A., Ward, D., 2011. A systematic review of interventions for promoting active transportation to school. International J. of Behavioral Nutrition and Physical Activity 8, 10.

Doorley, R., Pakrashi, V., Ghosh, B., 2015. Quantifying the health impacts of active travel: assessment of methodologies. Transp. Rev. 35, 1–24.

Faulkner, G.E.J., Buliung, R.N., Flora, P.K., Fusco, C., 2009. Active school transport, physical activity levels and body weight of children and youth: a systematic review. Preventive Medicine 48, 3–8.

Kurka, J.M., Adams, M.A., Geremia, C., Zhu, W., Cain, K.L., Conway, T.L., Sallis, J.F., 2016. Comparison of field and online observations for measuring land uses using the Microscale Audit of Pedestrian Streetscapes (MAPS). J. Transport & Health 3, 278–286.

Maghelal, P.K., Capp, C.J., 2011. Walkability: a review of existing pedestrian indices. Urisa J. 23, 5–19.

Moura, F., Cambra, P., Goncalves, A.B., 2017. Measuring walkability for distinct pedestrian groups with a participatory assessment method: a case study in Lisbon. Landsc. Urban Plan. 157, 282–296.

Mueller, N., Rojas-Rueda, D., Cole-Hunter, T., et al., 2015. Health impact assessment of active transportation: a systematic review. Prev. Med. 76, 103–114.

van Loon, J., Frank, L., 2011. Urban form relationships with youth physical activity: implications for research and practice. Journal of Planning Literature 26, 280–308.

Sustainable and resilient urban spaces

Pedestrians, Urban Spaces and Health – Tira, Pezzagno & Richiedei (eds)
© 2020 Taylor & Francis Group, London, ISBN 978-0-367-46171-3

Improving city resilience through demand and supply urban dynamics

D. Chondrogianni
Ph.D. Candidate, Department of Civil Engineering, University of Patras, Patras, Greece

Y.J. Stephanedes
Professor, Director of ITS Program, Department of Civil Engineering, University of Patras, Patras, Greece

ABSTRACT: This research is part of a PhD study that aims to evaluate the expected effectiveness of urban planning methods and policies based on their contribution to city resilience. Toward this goal, this research aims to identify the dynamic causal relationships between demand and supply in the use of urban spaces. On the supply side, the structure and dynamics of the processes that decision makers follow is a main subject of the research. Based on the process dynamics, the research identifies the main performance indicators that decision makers, such as local stakeholders and municipal actors should monitor in order to evaluate urban regeneration master plans and select the optimal one. On the demand side, the key causal factors and elements that attract or displease citizens in urban public spaces are identified. The case study focuses on urban Patras coastal hub, which has been a major redesign topic over the years.

Keywords: urban planning, resilience, decision making, future cities, demand and supply dynamics

1 INTRODUCTION/STATE OF THE ART

With half of the world's population living in urban areas (WHO, 2014), addressing and mitigating the impact of phenomena that are associated with climate change on urban processes is critical. The future of urban areas can be designed based on a dynamic system framework, including global relations, local powers and emerging trends and needs in cities, e.g. smart cities (Schmitt, 2015). Identification of urban planning needs are key to improving the effectiveness of construction and management policies in urban areas and require quick and risk averse decisions, the results of which can only be evident in the future.

The urbanization models of the 20th century were often applied without taking into full consideration the future results and impact of their implementation. In the 21st century, the focus on competitive and pioneering cities targets urbanization plans that are manageable and shaped with main objective the maximization of cities' benefits in terms of economic and environmental impact, prosperity, and quality of life.

The approach of cities as complex dynamic systems, which are called to pursue new prospects and alternative development plans to make their future sustainable, seems necessary (Forrester, 1975). In European cities, risk management often addresses out-of-control exogenous environmental agents, and the ambition of moving beyond sustainable growth is reflected in the development of Smart Cities dynamic strategies that support their digital and social transformation.

Cities that operate as European Hubs, participating in TEN-T networks, are first to address these challenges. In addition, as their response is of high interest for the future of Europe, further exploration of their urban dynamics that are subject to rapid change and are called upon to adopt best practices of urban sustainability and smart development are crucial.

2 OBJECTIVE OF THE PAPER

The concept of city resilience has been directly linked to adaptation to climate change (Leichenko, 2011). The EU's objective of adapting to climate change (European Commission, 2013) refers to supporting progress towards a "climate-proof Europe". In this effort, the analysis and evaluation of the various methods of urban planning is crucial for the future of cities.

The main objective of this research is to propose a methodological framework for analysing urban planning and to approach cities as global dynamic systems, in which a range of developing risks should be managed, and new challenges should be met, subject to the time-dependent needs and priorities of a smart city.

In this paper, the main priorities of this framework are presented and the demand and supply urban dynamics are set in the center of the research. On the supply side, the structure and dynamics of the processes that decision makers follow is a main subject of the paper. Based on the process dynamics, the paper aims to identify the main performance indicators that decision makers, such as local stakeholders and municipal actors monitor in order to evaluate urban regeneration master plans and select the optimal one. On the demand side, the causal factors and elements that attract or displease citizens in an urban public space are identified, given that citizens' will to visit and "own" a public space has a critical impact on the main indicators used by decision makers.

3 METHODOLOGICAL APPROACH OF THE RESEARCH

The following methodological and research tools are selected to achieve the objectives of this reseacrh (Figure 1):

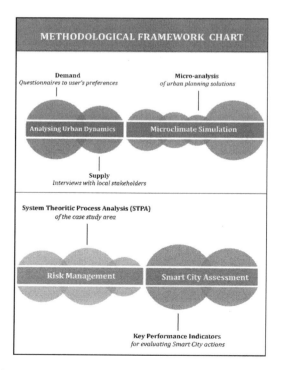

Figure 1.   Methodological framework diagram.

- Development of models for proposed urban design solutions using ENVI-met software, a tool for bioclimatic simulation and micro-analysis of urban design. The models will provide the opportunity to simulate the microclimate of the study area so that the effects of urban design proposals are measurable.
- Using the system theory tool, STAMP and the STPA risk analysis tool (Levenson, 2012) for risk management in the study area.
- Development of Key Performance Indicators of the proposed solutions for the study area in terms of their alignment with the Smart City Strategy of the urban area, and the actions of the EU Digital Cities Challenge program.

The European Hub of Patras is selected as the case study and, in particular, the urban area of the coastal front, for years a complex design issue, widely debated and revisited (e.g. Architectural Competition "Rehabilitation of Old Port of Patra", 2016). Addressing the fragmented approach of analysis of the area and the lack of scientific assessment of the proposed urban design solutions in terms of climate change, integration of "smart city" strategies and risk management is crucial for improving the citizens' quality of life and supporting Patras in becoming a growth model for Greek and European cities with common characteristics.

As a base for the research framework, the demand and supply relationships for the case study are identified and presented in this paper. The methodology for the identification of demand and supply causal relationships and parameters in Patras port city included deep interviews and discussions with local stakeholders and experts as well as questionnaires of users' revealed and stated preferences for urban spaces.

4 RESULT OF THE RESEACH

4.1 Citizens' preferences

For analysing the demand side in the urban spaces in Patras, a set of questions has been developed; 150 citizens responded by identifying the places they visit the most and the reasons for their preference, their main inefficiences and their impact on citizens' daily life. Among them, the 5 most important factors that determine citizens' decision to visit and remain in an urban space are the safety level, green spaces existence, air quality, noise pollution, and the existence of upgraded connected infrastructure (Figure 2).

In addition, research results revealed that citizens are aware of the importance of green spaces and the enviromental friendly infastructure in mitigating the negative effects of climate change and enhancing city resilience to extreme weather phenomena; they are less familiar with smart cities' aspects and the emerging implementation of smart cities strategies. However,

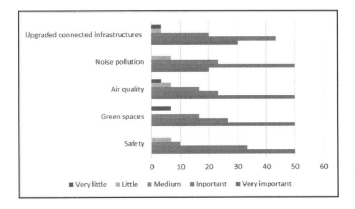

Figure 2. Percentage of the most important impacts on citizens' preferences for urban spaces.

they are positive to the incorporation and use of digital tools in the planning and regenaration of urban spaces and, in particular, for Patras seaside area (Figure 3). Their rating of specific smart applications for the study area are presented in Figure 4.

4.3 Decision making processes

From the interviews with decision makers, there was evidence for lack of a concrete process for urban regeneration decisions; such decisions almost entirely depend on the existence of the appropriate financing tools. In their opinion, monitoring the changing urban state through KPIs is critical; among these, *"citizens' demand to visit an urban space"* is the most indicative. Regarding the process of selecting an implementation plan, it is stated that,

> "There are no fixed criteria and the decision process depends on the criteria that each architectural contest or municipal stakeholders declare. Unfortunately, most criteria are based on the economic competitiveness of an offer and not on the technical excellence of that offer."

The majority of stakeholders is positive on the need for developing a methodological framework for assessing urban solutions for regeneration plans. They rank *green spaces, mobility* and the *upgraded connected infrastructure* as the main factors that should be considered in the decision making process. Regarding the case study area, the factors, *"upgrading the natural environment"* and *"implementation of smart applications"* are proposed as being of high priority for improving Patras' resilience and enhancing its capacity to respond to upcoming challenges.

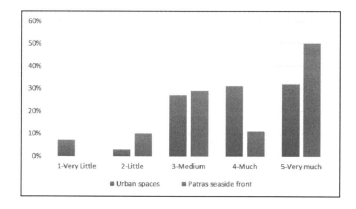

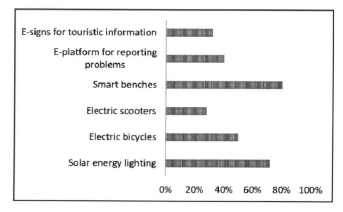

Figures 3 & 4.   Citizens' acceptance of Smart tools.

5 CONCLUSIONS

The complexity of the subject under study results in the need for addressing a range of barriers. The benefits from the holistic approach of modelling the cities' processes with emphasis on urban dynamcs will be both scientific and socio-economic. The scientific benefit lies in the innovation of the multidisciplinary analysis of a crucial, multilevel subject, combining various fields of research (environmental science, urban planning, risk management, decision making, systems dynamics). The socio-economic benefit derives from the development of a model of urban planning aiming to ensure urban resilience and prosperity.

Future research involves the assessment of urban design methods that affect the microclimate of urban area, the planning of a smart area on the coastal front of Patras that aims to improve microclimate and citizens'quality of life, and the development of risk analysis plans that are based on elaborate analysis of dynamic scenarios and hazards identification.

REFERENCES

European Commission, 2008, EU plan on Climate Change & The Covenant of Mayors.
European Commission, 2018, EU Digital Cities Challenge, Brussels.
Forrester, J., 1975, New Perspectives on Economic Growth, Alternatives to growth, A search for Sustainable Futures, Ballinger Publishing Company, Cambridge, Massachusetts, p. 107-121.
Huttner, S., Bruse, M., Dostal, P., 2010, Numerical Modeling of the urban climate – A preview on ENVI-met 4.0, Johannes-Gutenberg Universtpdelity Mainz, Germany.
Leichenko, R., 2011, Climate change and urban resilience, Current Opinion in Environmental Sustainability, Volume 3, Issue 3, p. 164-168.
Leveson, N., 2002, System safety engineering: back to the future, Massachusetts Institute of Technology.
Schmitt, G., 2015, Information Cities, the Future Cities Laboratory, Singapore and Zurich.
WHO, 2014, Report on Global Health Observatory (GHO) data, Urban health, World Health Organisation.

Urban resilience and anthropic risks. The case of the Polcevera Valley in Genoa

S. Candia, F. Pirlone, I. Spadaro & A.C. Taramasso
DICCA - University of Genoa, Italy

ABSTRACT: The research shows a methodological approach aimed at making urban spaces more resilient towards an external risk. To define the resilience of an urban system/space, it is necessary to assess the pre-event and after-event status. The resilience level is calculated on the basis of these two values. This is possible by dividing the system into macro-sectors and actors involved. In the literature there are several examples to evaluate urban resilience. This paper attempts to develop a resilience index using an actors/actions analysis. Indeed, it is possible to increase the threshold of resilience of the perturbed system by acting on each macro-sector, as a result of the implementation of a specific action by a determined actor. The authors consider as actors the components of the 'Quadruple Helix': universities, industry, governement and civil society. This methodological approach is applied to a case study: the Polcevera Valley (Genoa) after the Morandi Bridge collapse.

1 INTRODUCTION/STATE OF THE ART

The research shows a methodological approach aimed at making urban spaces more resilient towards an external risk. The first estimation done after a disaster (natural and anthropic) is the evaluation of lost lives, injures and the amount of direct damage relevant to the economical aspect. The variables utilized in this usual estimation are not influenced by the real state of the area affected by the disaster (damaged system in the following) and the previous state of the technical, economical, cultural and social conditions, that are different on the world, are not take into account on the classical estimation of damage. The previous state of the area and its effective functionality is easily represented considering the vulnerability and the resilience of the system damaged. The US National Accademy of Sciences deines resilience as "the ability to prepare and plant for absorb, recover from, and more successfully adapt to adverse events" (National Research Council, 2012). In literature (Abdulkareen et Elkadi, 2018, Marin et Modica 2017) not always it is possible to restore the previous functionality, and the time to come back to an acceptable level of the damaged system for the inhabitants and for the economical and social aspects of daily life is not always estimable.

2 OBJECTIVE OF THE PAPER

In particular, the paper, unlike the main experiences in the literature, deals with anthropic risks. According to the authors, the term resilience refers to a set of features that guarantees the maintenance of the territorial functionality and identity, as a result of external solicitations.

An event is often analysed considering only its direct effects; rarely the indirect effects on the surrounding area are investigated. It is not easy to define damaged system involved in the disaster analysis, and to identify the indirect impacts on the inhabitants' quality of life and the social, economic, and environmental aspects. It is important to outline different macrosectors in which it is possible to measure some parameters correlated to the different aspects of the resilience in the system to obtain a complessive evaluation pre-event and sucessive evaluations according to the different solutions adopted inside the damaged system.

3 METHODOLOGICAL APPROACH

To define the resilience of an urban system/space, it is necessary to assess the pre-event (time zero) and after-event functional status. The resilience level is calculated on the basis of these two values. This is possible by dividing the system into macro-sectors and actors involved. A variety of qualitative explanations are presented in the literature to analyze and address resiliency and vulnerability (Okine et al., 2009). This paper attempts to develop a resilience index for urban infrastructure using an actors/actions analysis. Indeed, it is possible to increase the threshold of resilience of the damaged system by acting on each macro sector, as a result of the implementation of a specific action by a determined actor. Most definitions of urban resilience are rather vague with respect to what constitutes an urban area or city (Merrow et al., 2016). For this reason, the authors defined different macro-sectors to describe and characterize the urban system under analysis: transport, education, health, public spaces, environmental and socio-economic framewok. These sectors are determined taking into account the EU Cohesion Policy 2014-2020 which identifies accessibility to transport, health and education as primary conditions for every European citizen. It is possible to increase the threshold of resilience of the perturbed system by acting on individual macro, as a result of the implementation of specific actions. These actions involve different actors. In fact, the identification of resilience strategies requires the simultaneous involvement of different stakeholders. This approach, called the Quadruple Helix model of innovation, focus on the urban space considered as a platform for interactions among different players. According to this model University, Entreprises, Public Authorities and Citizens work together to better align the innovation process with common goals and expectations. An urban system can react in a positive manner to a disaster if there is a synergy among these four actors, able to increase the macrosectors resilience (in terms of functionality and recognition). According to the methodology proposed by the authors, to calculate the level of resilience before and after the crisis, it is necessary to evaluate the actions and impacts of each of the actors of the Quadruple Helix on the macrosectors. These impacts can be positive, negative or neutral, depending on whether the actors involved have carried out effective or ineffective actions or have decided not to act (Table 1).

Table 1. Actors/actions matrix to analyse the resilience index for urban area.

Macro-sectors	Actors involved	Actors'actions/impacts
Transport	Public Authorities	Positive/Neutral/Negative
	Universities	Positive/Neutral/Negative
	Entreprises	Positive/Neutral/Negative
	Citizens	Positive/Neutral/Negative
Education	Public Authorities	Positive/Neutral/Negative
	Universities	Positive/Neutral/Negative
	Entreprises	Positive/Neutral/Negative
	Citizens	Positive/Neutral/Negative
Health	Public Authorities	Positive/Neutral/Negative
	Universities	Positive/Neutral/Negative
	Entreprises	Positive/Neutral/Negative
	Citizens	Positive/Neutral/Negative
Public spaces	Public Authorities	Positive/Neutral/Negative
	Universities	Positive/Neutral/Negative
	Entreprises	Positive/Neutral/Negative
	Citizens	Positive/Neutral/Negative
Environmental and socio-economic framewok	Public Authorities	Positive/Neutral/Negative
	Universities	Positive/Neutral/Negative
	Entreprises	Positive/Neutral/Negative
	Citizens	Positive/Neutral/Negative

Urban systems indeed are the result of the interactione between the natural and morphological aspects with the different actors who live, work, study or govern it. To quantify the resilience of the urban system a quali-quantitative analysis has been developed. In particular, positive effects are measured with values from 1 to 3, negative effects from -1 to -3, while the neutral effects are associated with the value 0.

4 DESCRIPTION

4.1 *Innovative aspects*

In the 2000s the method of Triple Helix developed involving institutions, industry and universities. Today it is necessary to move on to the concept of Quadruple Helix that involves also the end users of governance policies, the citizenship. According to the Quadruple Helix government, industry, universities and civil society bring multidisciplinary viewpoints together promoting the sharing of ideas. The methodology proposed by the authors is innovative in that it combines the concept of resilience to that of the Quadruple Helix for smart development. This approach can create new joint value that benefits all participants in what becomes an innovation and resilient ecosystem. This model shifts attention to the territory, which becomes a platform for relations among the actors involved (Maini, Nicolli, 2011). According to the authors, only through the involvement and activation of all four members of the Quadruple Helix can an urban system react positively with respect to external negative events and thus prove to be resilient.

4.2 *Research and/or technical developments*

The paper reports a concrete application of the methodology proposed by the authors to a case study: the Polcevera Valley after the Morandi Bridge collapse (14th August 2018). This is an urban area within the city of Genoa with about 61,000 residents and with more than 2,000 commercial activities. This negative event caused direct and indirect impacts, usually estimated in economic terms only. While the direct impacts are related to the loss of human lives and the interruption of an important arterial road for Genoa, the indirect impacts concern various aspects according to the scale of the urban system involved (all the valley and/or the neighborhood under the bridge). The Polcevera Valley has been affected by structural and functional impacts on its buildings, economic activities, public/private services, social behaviours and mobility. More generally, it can be stated that the entire city of Genoa, after this negative event, is having repercussions on its tourist, economic, infrastructural, social and environmental features, decreasing the quality of life of its citizens.

Following the methodology described in the previous chapter, the authors evaluated the resilience index of this area starting from the actions undertaken by each of the four key actors with respect to the individued urban macro-sectors. This index is measured before and after the collapse of the bridge. This means that the authors have assessed the actions done by the university, the municipality, the inhabitants and the industry/economic activities to improve: the mobility, health and education accessibility; the design of public spaces and the environmental- socio-economic framewok.

5 RESULT OF THE RESEACH

This section gives an example of the results obtained by applying the methodology described above to the first macro-sector: the transport system. The authors analyzed and evaluated the impacts related to the actions carried out by each actors of the Quadruple Helix in relation to urban mobility in Polcevera Valley. The table below lists all the actions undertaken by the municipality, the university, the busines sector and the citizens following the collapse of the Morandi bridge (Table II). The authors linked each action to a score based on its effectiveness. About Public Authorities, for example, the extension of the underground opening hours made it possible for citizens to travel at night (very positive impact: 3 points); the launch of a new free bus line did not have a significant impact on the local public transport network (neutral impact: 0 points) and the introduction of a new train timetable simply replaced the winter timetable with the summer timetable two weeks earlier without adding the new services needed (negative impacts: -2points). The results obtained (Table 2) have been reported on a radar graph suitable to highlight the contribution in terms of resilience of each actor involved before and after the negative event.

The same methodology has been-applied to all other macro-sectors giving a complete methodological framework of the level of resilience of the urban area under analysis.

Table 2. Extract of actors/actions matrix to resilience index for the transport.

Actors involved	Actions after the collapse of the bridge	Score
Public Authorities	Small interventions on the local traffic	2
	Extension of the opening hours of the metro	3
	New free bus line	0
	New train schedules	-2
	Sustainable Urban Mobility Plan -SUMP- definition	2
	Shuttles for students	1
	New connection with the highway	2
	Rethinking School Hours (flexible hours) and educational calendars	-1
	Fundraiser	2
	. . .	
Universities	Participation in the SUMP definition	2
	Mobility analysis through theses and workshop	2
	Smart Working for professors and employees	2
	Support for students residing in Polcevera Valley	1
	PRINCE project to support sustainable mobility	1
	Fundraiser	2
	. . .	
Entreprises	Promotion of telework	2
	Corporate car sharing	1
	Change of habits for freight	2
	Timetable policies	2
	. . .	
Citizens	Shift towards sustainable mobility, limitation of car use and use of local public transport (bus, train, metro, pedestrian . . .)	3
	Changes of habits in daily life (not only in terms of transport but in actions through-out the day: early wake-up, late evening returns, limitation in common activities, ..)	3
	Extension of travel times and exhausting journeys (few services against supply and poor quality)	3
	Actions of collaboration between citizens (pedibus for schools, special self-driven shuttles, . . .)	3
	Public commitment and desire to get up again	3
	Fundraiser	2
	Different approach among citizens of the same city (the citizens of eastern Genoa - not interested - not very participatory)	-2
	. . .	

Macro-sector: TRANSPORT
━━━ BEFORE ━━━ AFTER
Actors' impacts before and after the collapse of the bridge

6 CONCLUSIONS

The approach proposed in the paper shows how the actions developed by the sectors of the Quadruple Helix affect the overall resilience index of an area affected by an event. In the case study, the behavior of citizens was decisive, very participatory and active, which led to increasing the resilience of a sadly devastated area by an 'anthropic' event.

REFERENCES

Maini E., Nicolli F., (2011) Le politiche Europee per l'innovazione, in Poma L., Nicolli F. (a cura di), Franco Angeli, Milano.

Marin G, Modica M., (2017). "Socio-economic exposure to natural disasters", Environmental Impact of Assessment Review,Vol. 64, p.57–66.

Merrow S., Newell P., Stults M. (2016). "Defining urban resilience: A review", Landscape and Urban Planning, Vol. 147, March, p. 38–49 National Research Council (2012). Disaster resilience: a national imperative. The US National Accademy Press.

Okine N., Cooper A., Mensah A. (2009). "Formulation of Resilience Index of Urban Infrastructure Using Belief Functions", IEEE Systems Journal, Vol. 3:2, p. 147–153.

Pedestrians, Urban Spaces and Health – Tira, Pezzagno & Richiedei (eds)
© 2020 Taylor & Francis Group, London, ISBN 978-0-367-46171-3

Gamification for the enhancement of urban spaces: Cases and tools

M. Sciaccaluga & I. Delponte
Cieli, Università degli Studi di Genova, Italy

ABSTRACT: Real catastrophic events, such as the fall of Ponte Morandi in Genoa and alternative visions from the work of the contemporary artist Christo Yavachev, provide a starting point to reflect on how the citizen often has only partial knowledge of the territory with its potential and criticality. In a condition of necessity, a particular form of readjustment, known as resilience, leads the citizen to adapt to the modified environment, exploiting it synergistically and discovering it in a new and in-depth way. From this observation, the need for a different kind of interacting elements emerges: a tool capable of increasing the awareness of the city, the infrastructures and the urban landscape in citizens, the real users of spaces and public works. Participatory planning, in which the opinion of citizens has a role in the creation of policies and implementation projects, has more and more scope of action and intends to increase its work also using ICT to facilitate the process.

1 INTRODUCTION AND STATE OF THE ART

In situations of need, citizens tend to adapt their lifestyle according to the changes around them: the community demonstrates to have the resources to evolve and change coherently with the environment.

A great example of this is represented by Genoa, where the catastrophic fall of the Morandi motorway bridge has completely changed the life of the city and the mobility started to modify itself drastically. Administration, citizens and infrastructures, started to show a powerful resilience and began evolving, working as a unique living being. The administration created an efficient grid of communication with citizenship; the population started exploring new possible paths to move in the city and began to suggest them to friends and neighbours. As Christo's artistic work highlights, depriving people of a relevant and functional part of their city leads them to look at it with different eyes, to try new solutions and to learn new paths. (Contini Galleria d'Arte, 2015)

A reflection arises: would the mobility and the entire city system gain benefit if people had a deeper understanding of the territory? Is a catastrophic event or a work of art really necessary for this knowledge to arise? Is it possible to control it and favour systematic cooperation between citizens and with governance?

The need for this new type of community has made itself explicit and clear: including future citizens in the decisional process (Mukhtarov et al., 2018; Arnstein, 1969) can lead to a virtuous mutual exchange in which they can make informed decisions and the government can take advantage of the user information (Goodchild, 2007). Numerous frameworks have been proposed in literature, based on different kinds of ICT-facilitated systems of interaction between citizens and government, like the one developed by Farhad Mukhtarov et al. and based on the work of Linders (2012). To encourage civic engagement in urban planning many serious games (Social Impact Games, 2008) have been developed during the past years to support learning and fulfilment of specific duties in a playful and engaging way (Brabham, 2009; Juraschek et al., 2017).

A major issue is represented by rational ignorance: "...when the cost of educating oneself about the issue sufficiently to make an informed decision can outweigh any potential benefit one could reasonably expect to gain from that decision, and so it would be irrational to waste time doing so" (Krek, 2005, pag.2). Gaming experiences avoid this obstacle since directly living the space and moving in it is the best way to get into the flow reducing greatly the perceived effort (Gordon et al. 2011, page 508). The final aim is to lead citizens to collaborative rationality to understand collective problems and solve them (Habermans, 1981; Inees and Booher, 2010).

If the participatory process it's a collective aspect of the community, urban perception is an individual experience. It is important to distinguish the two components: to make a concrete participatory work it is necessary to work first on the individual's education and perceptive response. A holistic process of perception through participation is necessary to manage human-space relationships, considering the individual at different scales and using community and social networks as tools (Spiridon and Pacurar, 2016).

The collection and exchange of information can be helped and made more efficient thanks to ICT technology. The term "smart city" indicates a urban space able to use ICT to improve lifestyle, efficiency and competitiveness, in a sustainable and future-preserving way (International Telecommunication Union; Kondepudi, 2014). But it is under the definition of Smart City 2.0 that the individual perception, citizen centred design and citizens-inclusive governance come into play (Pomeroy, 2017).

2 OBJECTIVE OF THE PAPER

The article aims to give recommendations about the characteristics and features that an ICT tool should have in order to engage citizens in the participation matter. The authors present guidelines and tools for the design of a training tool able to test variations of the features and the corresponding variations in appropriate KPIs.

3 DESCRIPTION

Both traditional formats of participation and innovative thematic serious games contain useful elements able to create a consistent reference framework for participation methods. After an attentive analysis of the state of art of classic civic participations systems, urban ICT tools -consisting in traditional information exchange formats- and successful serious games about tactic urban themes, the authors propose suggestions, guidelines and features to design a digital tool for civic participation. User experience must be as natural and instinctive as possible. It's essential to guarantee a good trade-off between ease of use and guidance to the task desired. The aim of the tool is collecting data, involving and engaging the citizenship in understanding and shaping their urban spaces and laying the basis for a synergetic planning made with citizens (participatory planning) and one focused on the needs expressed by citizens (citizen cantered planning). Authors will also provide an evaluation framework to compare effectiveness and efficiency of such tools, enabling future studies to weigh up pros and cons of existing tools and use the result to design future technologies.

The main aspects analysed in the research are: graphic and aesthetic of the interface, visualization of the territory and of urban spaces, interaction system and territory shaping tasks and inputs, rewards and perceived goals, social aspect and community driven interactions as well as the data collection system and analysis.

The graphic appearance and the interface design are fundamental elements for a well-functioning and efficient tool. Every part of the graphic structure must be appealing and simple to understand, balancing aesthetic and functionality.

The visualization of the urban space is part of the core. It needs to be realistic -to allow the user to take vision of the urban system in a correct way- but also simplified in order to encourage the user to "play" without effort. In the last years the trends are 2D maps visualizations and augmented reality representation of space. Both can be realistic or codified.

In a tool for civic participation the user must interact with the territory, modifying it while expressing his own ideas and suggestions. The changing process is the result of creative expression of the user, but also the mean to highlight real needs, problems and suggestions for the urban space. While playing, the user must disclose his latent knowledge. Movement and drawing can help in the task. The type of movement must be coded and simply to actuate. The sliding method is usually well perceived by the user but in the last period, thanks to the spread of mobile games such as *Pokemon go*, many users are getting accustomed to the mixture of physical movement, augment reality and 2D representation all together. The changes can be performed trough the drag and drop of pre-selected elements, with pre-defined multi-choices menus, or via a more free expression method, like drawing, but the more expression power is given to the users, the more complex it will be to extract useful information from the collected data, so it is always important to reach the best compromise between freedom of interaction and objectiveness of the analysis.

As reward for the user many achievements can be unlocked, giving to the citizen virtual and in-app rewards. This feature helps to involve the user, increasing his retention. Another method, often used for action of civic participation to public causes, consists in giving physical rewards such as free tickets for events. The prize giving can occur according to a leaderboard that stimulate the competition and can become part of the social element of the game.

The creation of a network between citizens is a key point in the participation matter. The social aspect can be better visualized with the creation of a club system consisting in the aggregation of users operating in the same area or signalling the same issues in order to discuss and comparing ideas in a process of shared design. The social component should be developed also using social networks. Sharing badges, ideas and images of the outcome can enhance the information spread, tempting other potential users to start using the tool. Choosing the social networks (Facebook, Twitter, Instagram...) can change significantly the spread of the app, its result, the target age and retention.

Like in big data analysis, the approach to data collection must be as ecological as possible, therefore applying filters to information collected can help in the task: the user can play freely, having fun in changing the topography of the city, but it's possible to hide in the tasks and goals the issue object of designers' analysis, registering only the citizens' reactions to that specific matter. In this way the user will feel free to act according to his reflections while providing valuable data for the analysis to come. In addition, the tool could register issues about spatial elements that have been signalled or changed by more than a selected percentage of user, arising the interest of professionals. Collocated between traditional methods and serious games, this type of tool aims to fill the gap thanks to the gamification system, in order to overcome the rational ignorance of citizens, providing a fertile foreground for the analysis to come. Gamification is "a process of enhancing services with motivational affordances in order to invoke playful experiences and further behavioural outcomes" (Huotari and Hamari, 2012), its use can create a virtuous circle in which using citizens' knowledge to form citizens' consciousness. Following the idea of citizens as sensors, the data collection could take advantage of a widespread system of signalling ad suggestions.

Authors also want to suggest efficient ways to evaluate the efficacy of similar tools. Concerning the market of traditional mobile games, the most used method to evaluate the application consists of KPIs (Key Performance Indicators). KPIs help in calibrating the contents and metadata of the app in order to achieve the wanted goals.

The KPIs selected for a public participation app are the following: distribution by age groups, richness of information provided, retention, pleasantness of the experience, usability of aggregated data, communication skill toward the citizen, number of installations.

To optimize the functioning of the app, it's necessary to study and analyse the positive or negative variations of KPIs according to the variation of specific features of the app itself.

Every characteristic can be implemented in different ways, each of which may influence the related KPIs. To decide what way is the best for the application and its user, game industries use *the variation test or AB test*: the application is published in different versions, according to the variation of characteristic, and distributed to different percentages of users. Then the KPIs of the versions are analysed and compared to choose the most efficient variation and optimizing the app accordingly.

4 CONCLUSIONS

The authors have emphasized the importance of civic engagement and participation action. They have presented a list of key essential elements and highlighted necessary KPIs able to assess the efficacy and efficiency of participation app tools. They provide elements to create a framework on which basing the development of future public participation apps and tools for civic consciousness and awareness. A framework based on already mentioned KPIs, essential features and the results of AB tests will allow to compare the results of different geographical areas, underlining common points and differences to investigate. The obtained results can be the basis for future urban-social analysis.

4.1 *Barriers and drivers*

Authors presents suggestions and guidelines deriving from the state of art analysis and their previous work experiences. The suggested framework needs to be tested and validated and the development of a training application tools would be necessary. Moreover, the users profiling, divided into demographic categories, needs to be deepened in order to take motivated decisions on the percentage of users to involve in the different ab tests.

REFERENCES

Arnstein S., (1969). A ladder of citizen participation. Am. Ins. of planners.

Brabham D., (2009). Crowdsourcing the Public Participation Process for Planning Projects. Plan the Theory, 242–262.

Contini Galleria (2015). Biografia, Bibliografia, Esposizioni Christo.

Goodchild, M., (2007). Citizens as sensors: the world of volunteered geography. GeoJournal, 211–221.

Gordon E., Schirra S., Hollander J., (2011). Immersive planning: a conceptual model for designing public participation with new technologies. Environment and Planning, 505–519.

Habermas J., (1981). The theory of Communicative Action: Reason and the Rationalization of Society. Beacon Press.

Huotari K., Hamari J., (2012). Defining gamification: a service marketing perspective, 16th I.A. Mind-Trek Conference.

Innes J., Booher D., (2010). Planning with Complexity: An Introduction to Collaborative Rationality for Public Policy. Routledge.

Juraschek M., Herrmann C., Thiede S., (2017). Utilizing gaming technology for simulation of urban product. Elsevier.

Kondepudi S., (2014). Smart sustainable cities analysis of definitions. The ITU-T focus group.

Krek A., (2005). Rational ignorance of the citizens in public participatory planning. International Symposium Vienna, 557–564.

Linders D., (2012). From e-government to we-government: defining a typology for citizens' coproduction in the age of social media.

Münster S., Georgi C., Heijne K., Klamert K., Noenning J., Pump M., Stelzle B., Meer H., (2017). How to involve inhabitants in urban design planning by using digital tools? An overview of a state of the art, key challenges and promising approaches. Elsevier.

Mueller J., Lu H., Chirkin A., Klein B., Schmitt G., (2018). Citizen Design Science: A Strategy for crowd-creative urban design. Elsevier.

Mukhtarov F., Dieperink C., Driessen P., (2018). The influence of information and communication technologies on public participation in urban water governance: A review of place-based research. Elsevier.

Pomeroy J., (2017). Smart Cities 2.0. A 8-part tv series.

Poplin A., (2012). Playful participation in urban planning: A case study for online serious games. Elsevier.

Poplin A., (2014). Digital serious game for urban planning: "B3-Design your Marketplace!". Planning and Design.

Social Impact Games, (2008). http://www.socialimpactgames.com.

Spiridon L., Păcurar L., (2016). Urban perception: A study on how Communities' Necessities regarding the City's Form and Function have changed throughout Time. The New ARCH.

Nature-based solutions for urban resilience

R. De Lotto
Department of Civil Engineering and Architecture, University of Pavia, Italy

F. Pinto
Department of Architecture and Urban Studies, Politecnico di Milano, Italy

ABSTRACT: The aim of the paper is to highlight the connections among the different applications of Nature based Solutions to enhance the capacity of an urban system to be more and more resilient. The study starts from the analysis of the different dimensions of resilience, that lately assumed a central role in the management of urban actions towards sustainability. Then some parameters are defined, based on the specific objectives of sustainability and a set of indicators is helpful to try to measure how much a specific urban action works towards a resilient behaviour. On the other hand, Nature based Solutions emerged as new keyword to define different ecological actions and tools: from the so called "green infrastructures" to the local application of building scale solutions (green roofs, pocket gardens, etc.). It is interesting to investigate how Resilience and NBS interact on the local urban scale.

1 INTRODUCTION

In recent years there has been a proliferation of studies on improving the resilience of cities against a multitude of artificial and natural disasters. There has also been an increase in the number of parameters and indicators developed to assess urban resilience. As climate change advances, resilience will become an increasingly significant topic in science and politics and will certainly influence future urban development. In particular, the resilience indicators are fundamental to support planners and decision makers to understand the fragility of their communities and develop strategies and action plans to create more resilient cities. Urban areas have had to strengthen their efforts to face the increasingly widespread effects of climate change. A new way to face environmental challenges and think about the future of cities are the so-called "nature based" solutions, such as parks, green areas, facades and green roofs for buildings. These solutions can be also economically advantageous and, at the same time, capable of generating benefits for the environment, the quality of life and the economy. The report "Nature-based solutions to promote climate resilience in urban areas - developing an impact evaluation framework" carried out within the EKLIPSE project (EU Horizon 2020 programme), offers a complete picture on the possible benefits of blue and green solutions and a support to find strategies and measures with the aim of assessing costs and benefits. Specifically, some solutions have been identified to meet ten important challenges for European cities: resilience to climate change, water management, defense of coastal areas, green space management, air quality, the urban regeneration, the relationships between physical urban space and the well-being of citizens, social justice and social cohesion, the creation of new economic opportunities. There is a growing demand for strategies aimed at providing the city with its resilience capacity. Therefore, it is necessary to identify the necessary policies so that the themes of territorial risks are addressed with concrete planning actions and become fundamental strategies for the adaptive and resilient city project.

2 ACTIONS AND PARAMETERS FOR THE EVALUATION OF URBAN RESILIENCE

Resilience is a concept that presents various definitions, probably due to the fact that it has been adopted by various disciplines that have interpreted it differently according to their needs and priorities. As cities are socio-ecological systems, which present dynamic interactions through time and space, the adaptive approach to resilience can provide the most suitable theoretical basis for conceptualizing urban resilience. This approach is reflected in the definition of the National Academy of Science (NAS) as "the ability to successfully prepare and plan, absorb, recover and adapt to adverse events" (Committee on Increasing National Resilience to Hazards and Disasters 2012). To achieve, maintain and strengthen these capacities, each urban system should have a series of characteristics such as: A) robustness, B) stability, C) flexibility, D) resourcefulness, E) capacity for coordination, F) redundancy, G) forecasting, H) independence, I) connectivity, J) capacity for collaboration, K) agility, L) adaptability, M) self-organization, N) creativity, O) efficiency, P) equity (Sharifi and Yamagata 2016) – capital letters of the list are given by paper's authors for further utilization. When thinking of these characteristics it must be remembered that there are synergies and compromises between the various dimensions. In order to develop a global evaluation system, it is necessary to deepen these synergies and compromises and to clarify how each of the characteristics is related to urban planning. Resilience is characterized by multiple dimensions that should be analyzed in an overall assessment framework of the city system. The study analyzes various criteria in the literature that can be used for the development of a resilience assessment system. After a theoretical-methodological study conducted through the analysis of urban plans, sharing platforms, guidelines, and the filing of numerous case studies, it is possible to define intervention methods, which can be implemented both in critical conditions and in the phase of design as a preventive action, if you want to apply the concept of resilience to climate change on buildings and neighborhoods. The first step concerns a collection of strategies and actions, grouped by sector or area of criticality, mitigation and adaptation to be undertaken to achieve urban resilience. The most commonly used methods of intervention are then presented, exposing their strengths and weaknesses and the best contexts of application, also indicating the time required for implementation, the useful life of the actions and the flexibility category. The intervention methods, grouped by macro areas, criticalities and actions are evaluated through a TOWS matrix (external factors: Threats, Opportunities, internal factors: Weaknesses, Strengths). Then two sets of parameters are defined, the first to be used to compare the adaptation and mitigation actions undertaken; the second represents an evaluation methodology, which bearing in mind the triple dimension, environmental social and economic, of resilience attributes a score, which in addition to allowing to distinguish actions of pure adaptation from resilient strategies allows to have an estimate of the validity of the strategy undertaken. Through the collection of case studies and the re-elaboration with grouping of the strategies to be followed, comparison parameters have been established, some of which can be evaluated quantitatively, while others are only qualitatively. These parameters can be used to evaluate the extent of the adaptations and, in order to allow a direct comparison between the actions taken, it is possible to obtain a dimensionless parameter, relating the extent of the intervention to a different scale. The proposed evaluation methodology is useful not only to understand the terms in which the measure undertaken can lead to the achievement of urban resilience, but also to have a trace to follow, to design a new neighborhood or even to recover an existing one. Each parameter represents a category that contains a series of actions, precisely because we want to highlight that it is not the action that defines resilience, but the objective we want to achieve. The objectives can be achieved in several ways, it is not necessary that all the factors are present to consider a resilient approach, one must always bear in mind the problem to be addressed and solve it. The evaluation can be performed on the intervention guidelines, at the building, neighborhood or city level with the appropriate assessments to be made on a case by case basis.

3 RESEARCH PROPOSAL FOR NATURE BASED SOLUTIONS (NBS) CLASSIFICATION

Nature based Solutions come from an EU specifici Research and Innovation Policy agenda on "Nature-Based Solutions and Re-Naturing Cities aims to position the EU as leader in 'Innovating with nature' for more sustainable and resilient societies." (EU, Nature-Based Solutions) that was implemented in H2020 Topics and researches. The main goals were oriented to: Re-naturing cities and Territorial resilience. So the connection between NbS and Urban resilience comes directly from the definition of the issues. NbS are defined as "societal challenges as solutions that are inspired and supported by nature, which are cost-effective, simultaneously provide environmental, social and economic benefits and help build resilience" (EU, 2015). A specific H2020 research called "Nature4Cities" started in 2016, to which authors participated as consultants, classified NbS according to some strategies, objectives, forms, descriptive elements, parameters, enablers and inhibitors, and implementation models. The classification is made of a hirerchical scheme depending on specific cathegories defined according to the cultural and practical environment in which NbS are defined. This scheme is not exclusive but inclusive. It means that different carachteristics that belong to the same chategory, may be tagged in the same time. Or, moreover, that characteristics belonging to different categories can appear simultanuosly. This fact depends on the specific nature of NbS: the definition itself of NbS is a "framework definition", in which many different examples, case studies and experiences are involved (De Lotto, 2017). In Nature4Cities database, for each example they are defined: general description, urban challenges addressed, description of the nbs, governance model, financing mechanism, nbs business model, legal aspects, temporal perspective, success and limiting factors of the intervention. Of course, the last two elements are extremely relevant considering the effectiveness of NbS and the possibility to register the promised results in certain time. But for the main aim of the paper, authors selected the "Urban Challenges Addressed" as the category that mainly may coordinate the selection of NbS as possible drivers to enhance Urban Resilience. Among the urban challenges in the classifications there are: 1. climate issues, 2. Urban water management and quality, 3. Air quality, 4. Urban space and Biodiversity, 5. Urban Regeneration and Soil, 6. Resource efficiency, 7. Public health and wellbeing, 8. Environmental justice and social cohesion, 9. Urban planning and governance, 10. People security, 11. Green economy, 12. Others.

4 PROPOSED METHODOLOGY: QUALITATIVE CORRELATION MATRIX

Once both Urban Resielince and NbS have been defined and described, the next step is to analyze if and how much each urban challenge of NbS is able to be a driver for Relience goals. Many complex connections exist but the general behaviour can be synthesized with a qualitative "correlation matrix" underlining strong, average and weak connections among variables. In the next table, for what references stated, it is generally excluded that two variables are not influencing each other. It means that each NbS may have a positive effect in urban resilience, like references underline. So, author considered to attribute qualitative values among three values of interrelation: weak (W), medium (M), strong (S). In Table 1 there is the synthesis of the correlation matrix (excluding the voice: 12. Others).

5 COMMENTS AND NEXT STEPS OF THE RESEARCH

Among the 176 interactions, there is a prevalence of strong relations (108, 61%) while average (31, 18%) and weak (38, 21%) are highly lower. This basic result is quite predictable, as well as the NbS categories "Urban planning and Governance" and "Green Economy" are the ones who demonstrate the strongest effectiveness toward resilience. The general overview demonstrates that almost 20% of the correlation are weak: it could mean that a specific analysis must be deepened considering specific characteristics of NbS and of the specific resilience goals. Reading the matrix by rows, the goals K) agility, M) self-organization and N) creativity have a limited number of

Table 1. Correlation matrix.

NbS Urban challenges driving Urban Resilience characteristics	1	2	3	4	5	6	7	8	9	10	11
A	S	S	S	S	W	S	W	S	S	W	S
B	S	M	S	S	S	S	M	S	S	S	S
C	S	S	M	S	S	M	S	S	S	W	S
D	S	S	M	S	S	S	S	M	S	W	S
E	M	W	W	S	M	S	S	S	S	W	S
F	W	M	W	S	M	S	W	W	S	M	S
G	S	M	M	S	S	S	W	W	S	S	S
H	M	S	S	M	W	W	S	S	S	S	S
I	S	S	S	S	M	M	S	W	S	M	S
J	S	M	S	M	W	M	S	S	S	M	S
K	W	W	W	M	W	M	W	S	S	W	S
L	S	S	S	S	S	S	M	S	S	W	S
M	S	W	W	S	W	W	W	S	S	M	S
N	M	W	W	S	S	W	W	W	S	S	S
O	S	S	S	S	S	S	S	W	S	M	S
P	W	M	M	S	M	S	S	S	S	S	S

clear NbS drivers. This depends on the pure strategic but precise nature of these parameters in comparison with the general character of NbS classification. The next step that naturally emerges, is to concentrate the attention on specific NbS examples belonging different categories and analyze how much they are able to satisfy or drive urban resilience. So, starting from the previous research steps, and after a first view of the possible interactions emerging from general categories, it is necessary to come back to single and very representative case studies in order to clarify the role that these relatively new typologies of elements may have in theorical and practical urban planning actions. From the first general examination here presented, the path seems to be a prosperous field.

REFERENCES

AA.VV. (2015) Nature-Based Solutions & Re-Naturing Cities, Towards an EU Research and Innovation policy agenda for Final Report of the Horizon 2020.

AA.VV. (2018) Nature4Cities, https://www.nature4cities.eu.

Battarra R., Pinto F., Tremiterra M.R. (2018). "Indicators and Actions for the Smart and Sustainable City: a study on Italian Metropolitan Cities" in Papa R., Fistola R., Gargiulo (eds.) "Smart planning: Sustainability and Mobility in the Age of Change", Springer International Publishing. (pp. 83–108).

De Lotto R. (2017). Nature-based Solutions: new EU topic to renature cities. Urbanistica Informazioni, vol. 272 X giornata di Studio INU "Crisi e rinascita della città", p. 798, ISSN: 0392–5005.

De Lotto R., Esopi G., Sturla S. (2017). Sustainable policies to improve urban ecosystem resilience. International Journal of Sustainable Development and Planning, vol. 12, p. 780–788, ISSN: 1743–7601.

Galderisi A., Mazzeo G., Pinto F. (2016). "Cities dealing with energy issues and climate-related impacts: approaches, strategies and tools for a sustainable urban development", in Papa R., Fistola R. (eds.) Smart Energy in the Smart City. Urban Planning for a Sustainable Future, publishing series "Green Energy and Technology", Springer International Publishing. (pp. 199–217).

National Academies (U.S.). Committee on Increasing National Resilience to Hazards and Disasters, (2012). Disaster resilience: a national imperative, National Academies Press, Washington.

Saporiti G., Scudo G., Echave C. (2012). Strumenti di valutazione della resilienza urbana, in TeMa, vol.5, n. 2.

Sharifi A., Yamagata Y., (2016). "Urban Resilience Assessment: Multiple Dimensions, Criteria, and Indicators", in Yamagata Y., Maruyama H. (eds.) Urban Resilience: A Transformative Approach. Springer International Publishing (pp.259–276).

Definition of pedestrian friendly street parameters and evaluation in the case of Erzurum city

E.N. Sari
Faculty of Forest, Istanbul-Cerrahpasa University, İstanbul, Turkey

S. Yilmaz
Faculty of Architecture and Design, Ataturk University, Erzurum, Turkey

B.G. Yilmaz
Architecture Faculty, İstanbul Kültür University, İstanbul, Turkey

ABSTRACT: In livable cities; the street is defined as the first encounter area when going from private to public space. It is the core area that will contribute to increase the quality of life in the city. In this sense, the street must meet the basic criteria for providing a comfortable living environment to urban users. In the scope of the study, an attempt was made to define pedestrian-friendly street parameters including the air quality and thermal comfort. Therefore, a microclimate data of the city center streets were hourly collected 1.5 m above the ground level in winter period of 2017. Then a different landscape design scenarios in pedestrian streets was investigated widely using **Envi-met** model to determine the thermal comfort level by measuring the predicted mean vote index (**PMV**). The result shows that, the scenario of deciduous trees was found to have higher thermal comfort than treeless street design for pedestrian-friendly streets. The study aims to fill the gap in the literature by taking the street ecology which is neglected in the user-friendly street works and by showing the pedestrian-friendly street parameters that are physically and ecologically meaningful This result coincides with the physical parameters and necessitates the revision of the vegetation arrangement once more. When thermal comfort and the air quality are taken into account, it is understood that streching the principles is a necessity in order to obtain a sustainable street design. The outputs of this study will help to improve the outdoor thermal comfort in first stage of the urban planning and landscape street design for more livable effective city.

Keywords: ENVI-met, CFD, livable street, wind, air qality

1 INTRODUCTION

Erzurum, a winter city with a continental climate at an altitude of 1800 meters approximately, is situated in Turkey's North East. An intensified air pollution during the winter season most particularly affects the standard of living unfavourably. A negative relation is identified between the inhabitability of the city and the air pollution (Martínez-Bravo et al. 2019). There are various parameters that affect the sustainability and the life quality in the cities. The environment is subcategorized under the headings of economy and equality in the literature (El-Shimy and Ragheb 2017; Martínez-Bravo et al. 2019). Under the topic of environment, the integrated approaches between sustainability and subscales (street scale) and upper scales (urban macroform) are assessed (Willams 2001; Badawi 2017; Bandarabad and Shahcheraghi 2012; Juaidi et al. 2019; Mahmoudi et al. 2015;Yilmaz et al. 2018; Reis et al. 2019; Sharifi 2019;; Yassin 2019). Studies in recent years proved that walkability is an essential element of the inhabitability that is supported by a sustainable environment and a liveable space (Yilmaz et al. 2014; Shamsuddin et al. 2012).A sustainable street design should embody certain principles such as diversity and choice, comfort and streetscape, safety, liveliness, environmental quality, and economic vitality. Street vegetation and a green infrastructure are also included to the principles that affect the walking comfort (El-Shimy and Ragheb 2017; Yilmaz et al. 2018a). In most studies, all these parameters are discussed separately and no basic comparative evaluation is made for urban design (Shashua-Bar and Hoffman 2000; Hartig and Kahn 2016; Janhäll 2015; Sarı, 2019). Vegetation cover has an impact on draining the rainwater, directing airflow, decreasing or increasing the air temperature, and providing shading effects.

Street vegetation can affect the intensiveness of the air pollutants in the streets by directing, accelerating or stemming the airflow straight-forwardly. On the other side, the thermal temperature of the streets also influences the standard of living by directly affecting the air quality.

2 OBJECTIVE OF THE PAPER

In this paper, determining the pedestrian-friendly street parameters for Erzurum, a characteristically winter city, is aimed by focusing on the Mumcu Street example which is located in the city centre where people actively attend to the urban life. In this context, the analysis of street with and without trees is carried out by focusing on the air quality and the thermal comfort in order to determine the impacts of the pedestrian-friendly street parameters on the environmental quality. In addition, the methods to provide the appropriate conditions in a sustainable street design for an ideal environmental quality are examined through the example of a winter city.

3 METHODOLOGICAL APPROACH

The measurable parameters and standards are investigated within the scope of this study. *In the beginning*, the qualities of Mumcu Street is identified by section cut analysis in order to assess the physical comfort of the pedestrians. *In the second stage*, ENVI-met is used for the purpose of analysing the thermal comfort of the pedestrians. Meteorological data were collected from the Mumcu Street in winter 2017, at an interval of 1 hour from a height of 1.5 m. 24-hour data were used for these measurements, while the remaining hourly data were kept waiting for the device's median optimization. This street area was analyzed for the current and proposed state respectively for winter. In this study time interval of 1 hour is used WS300 sensitive meteorology recorder was used in the collection of data. The latitude of this Mumcu Street is 39° 54′ 16.5816″N and the longitude is 41° 15′ 36.4896″E. While measuring the temperature values with the precision thermometer (WS300), the humidity values were taken with the YCOM-KMN 305 model temperature and humidity meter. TROTEC BA16 model anemometer is used for wind speed measurement. A three-dimensional micro-climatic 3D modelling tool ENVI-met was used to evaluate meteorological data. *In the third stage*, CFD is used for wind flow analysis with the intent of assessing the pedestrian comfort and the air quality. For each analysis, the forested and unforested situations are compared with each other. The inputs –the lowest temperature: -13,45 (06:00); the highest temperature: 8,7 (13:00); maximum wet: 95 (06:00); minimum wet: 48 (13:00); wind velocity: 2; wind direction: ESE (112,5°)– are processed through the programme. The analyses are carried out under the condition that the temperature is 0 °C and the wind velocity is 2 m/s by using ANSYS 16.0.

4 DESCRIPTION

4.1 *Study area location*

The street which is intensely used by the city-dwellers is situated in the city centre of Erzurum (Figure 1). The pedestrian access is provided only by pavements. The centre of the street is located in the working area. Along the street are business centers and cafes.

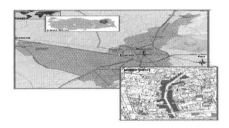

Figure 1. Location.

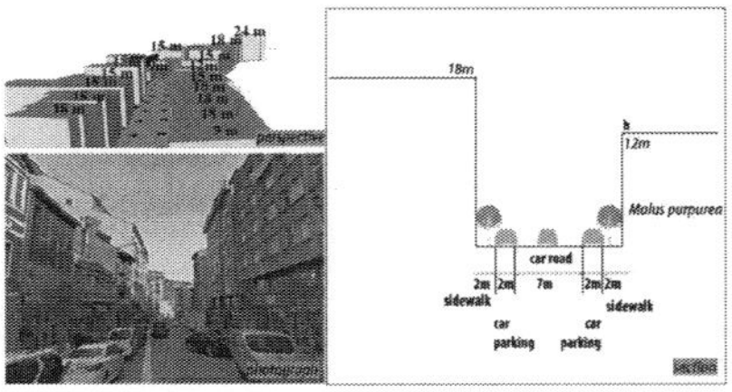

Figure 2. Street examination.

5 RESULT OF THE RESEARCH

5.1 *Analysis of ASYS16.0*

In order to estimate the air quality and street wind conditions, 3 different simulations were made: empty, tree and vehicle. As a result of the simulation, in the streets that exhibit canyon characteristics, the wind speed of the wind; Figure 3> Figure 4> Figure 5 respectively. Therefore, the removal of particulate matter is the same.

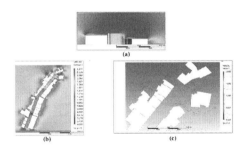

Figure 3. This analysis where the street is empty a)This analysis where the street is alley b) plan of street c) wind flow.

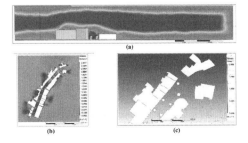

Figure 4. The analysis where the street is alley a)This analysis where the street is alley b) plan of street c) wind flow.

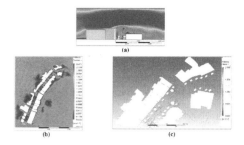

Figure 5.   This analysis where the street have tree and car a)section b) plan of street c) wind flow.

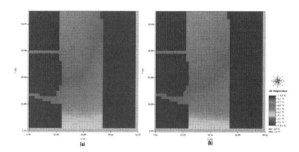

Figure 6.   A)This analysis where the street is Suggestion Analysis with trees b) a)This analysis where the street is no trees.

5.2   *Analysis of ENVI-met*

In order to forecast the thermal comfort of the street during the winter conditions, analyses are carried out in the streets both have trees and have not (Figure 6). The wind velocity and the direction of the wind remain constant during the analyses. Analysis of PMV is also made with the purpose of assessing the comfort impact of the changing temperature on the humans. The results of the analyses show that when the number of plants on the street is increased, the temperature increases in winters and decreases in summers (Yilmaz, et al. 2018a)

6   CONCLUSIONS

According to the results of the analyses, it is determined that the structural design qualities of the street influence the thermal comfort and the air quality directly. Initial portrayal of the street demonstrates that the physical conditions of the street does not actualize the all necessary conditions for pedestrian comfort. During the study performed with the aim of assessing the air quality, it is identified that the empty street is exposed to the wind more than the other simulations. When the wind velocity at roof and pedestrian level is examined, the apparent variations between the simulations are not investigated. This situation is associated to the existence of the upwind buildings, the upwind street direction and the tall buildings that tame the wind. No significant difference is found between the simulations at the end of the study conducted in order to evaluate the thermal comfort. No major contribution is also made to the improvement of the comfort of the environment in the thermal comfort analysis carried out under the winter conditions.

6.1 Barriers and drivers

Although the evaluation is carried out with the same parameters when assessing the air quality and the thermal comfort, the results do not meet the expectations. The necessary conditions cannot be provided to recover the air quality and it is also hard to assert that there are affirmative conditions for thermal comfort as well. Additionally, when assessed with the physical parameters, it is concluded that parked vehicles both influence the walking comfort in a negative way and also negatively affect the air quality by taming the wind velocity. It is impossible to make mention of a sustainable street design at this stage. On the other hand, when the analyses carried out for streets that have trees and those who have not are evaluated, it is confirmed that under the affirmative conditions, it has a neutral impact on the thermal comfort and a negative impact on the air quality. This result coincides with the physical parameters and necessitates the revision of the vegetation arrangement once more. When thermal comfort and the air quality are taken into account, it is understood that streching the principles is a necessity in order to obtain a sustainable street design.

The study emphasized that a multidisciplinary team should work together for establishing a healthy, sustainable and livable urbanization with street thermal comfort when planning the city.

ACKNOWLEDGEMENT

This study context used data of Elif Nur Sarı's master thesis and TUBITAK 1001 -2150627 number project derived. We thanks for ENVI-Met analysis to Başak Mutlu.

REFERENCES

Badawi, Samaa. 2017. "Sustainable Approach for Developing Local Mixed- Use Streets Case Study Beit Al Maqdis Street in Jeddah." Procedia Environmental Sciences 37: 374–85.

Bandarabad, Alireza, and Azadeh Shahcheraghi. 2012. "Livable Street in Urban Environment: An Adaptive Design Approach." Advances in Environmental Biology.

El-Shimy, Hisham Galal, and Riham Aly Ragheb. 2017. "Sustainable Urban Street Design: Evaluation of El-Moaz Street in Cairo, Egypt." Procedia Environmental Sciences 37: 689–98.

Hartig, Terry, and Peter H. Kahn. 2016. "Living in Cities, Naturally." Science.

Janhäll, Sara. 2015. "Review on Urban Vegetation and Particle Air Pollution - Deposition and Dispersion." Atmospheric Environment.

Juaidi, Adel, et al. 2019. "Urban Design to Achieving the Sustainable Energy of Residential Neighbourhoods in Arid Climate." Journal of Cleaner Production 228: 135–52. https://doi.org/10.1016/j. jclepro.2019.04.269.

Mahmoudi, Mohadeseh, Faizah Ahmad, and Bushra Abbasi. 2015. "Livable Streets: The Effects of Physical Problems on the Quality and Livability of Kuala Lumpur Streets." Cities 43: 104–14. http://dx. doi.org/10.1016/j.cities.2014.11.016.

Martínez-Bravo, María del Mar, Javier Martínez-del-Río, and Raquel Antolín-López. 2019. "Trade-Offs among Urban Sustainability, Pollution and Livability in European Cities." Journal of Cleaner Production 224: 651–60.

Reis, Inês F.C. et al. 2019. "An Evaluation Thermometer for Assessing City Sustainability and Livability." Sustainable Cities and Society 47(February).

Sari, E. N. 2019. " Analysis of the relationship between air pollution and housing pattern: Example of Erzurum ". Master Thesis.

Shamsuddin, Shuhana, Nur Rasyiqah Abu Hassan, and Siti Fatimah Ilani Bilyamin. 2012. "Walkable Environment in Increasing the Liveability of a City." Procedia-Social and Behavioral Sciences 50: 167–78.

Sharifi, Ayyoob. 2019. "Resilient Urban Forms: A Review of Literature on Streets and Street Networks." Building and Environment 147 (September 2018): 171–87. https://doi.org/10.1016/j. buildenv.2018.09.040.

Shashua-Bar, L., and M.E. Hoffman. 2000. "Vegetation as a Climatic Component in the Design of an Urban Street." Energy and Buildings.

Willams, Nick. 2001. "Achieving Sustainable Urban Form." Land Use Policy.

Yassin, Hend H. 2019. "Livable City: An Approach to Pedestrianization through Tactical Urbanism." Alexandria Engineering Journal 58(1): 251–59. https://doi.org/10.1016/j.aej.2019.02.005.

Yilmaz S., Yildız ND., Irmak MA., Yilmaz H., 2014. Effects of plant design on bioclimatic comfort in special areas; sidewalks. Third International Conference on Countermeasures to Urban Heat Island, October 13–15, 2014, pp. 592–601, Oral Presentation, Venezia, Italy.

Yilmaz S.,Mutlu E.,Yilmaz H., 2018. Quantification of thermal comfort based on different street orientation in winter months of urban city Dadaşkent. DOİ: 10.17660/Acta Horticultura, 1215.12, EdsG. Pennisi, L. Cremonini, T. Georgiadis, F. Orsini, G.P. Gianquinto, ISBN: 978-94-62612-12-9, ISSN: 0567–7572 (print) 2406–6168 (elect.), Acta Horticulturae, 1215: 67–72.

Yilmaz S., Mutlu E., Yilmaz H., 2018a. Alternative Scenarios For Ecological Urbanizations Using Envi-Met Model. Environmental Science and Pollution Research, 25 (26): 26307–26321.

Pedestrians, Urban Spaces and Health – Tira, Pezzagno & Richiedei (eds)
© 2020 Taylor & Francis Group, London, ISBN 978-0-367-46171-3

Well-being, perception, participation and mobility strategieS. The sustainability of the contemporary city.

M. Lisi, F. Fratini & A. Cappuccitti
Sapienza University of Rome, Italy

ABSTRACT: Transport is currently one of the major problems for the credibility of European governments towards international climate agreements (COP-21, UN-Habitat). The evolution of means has above all changed the life of the community and greatly influenced its socio-economic development. The costs of health and congestion, among other things, are draining billions from our economy every year (WHO). For these reasons a major commitment is required to get people out of the cars and to encourage active mobility. The achievement of these objectives presupposes a change of paradigm in the Planning Strategies based on the needs of the citizen, also thanks to the support of new information technologies. The empirical research will have a specific focus on the study of the PPGIS tools for the construction of surveys. The potential application of the results obtained will contribute to an implementation of methods and tools in participatory processes for research, planning or policy implementation.

1 INTRODUCTION

According to UN-Habitat, European cities face a number of environmental challenges, which affect the everyday lives of millions of citizens and, as the Paris summit (2015) highlighted, contribute to produce "unsustainability". The contemporary urban model based on consumption of fossil energy, land and greenery and on the massive use of private transport is highly unsustainable. Regenerating cities means improve the citizens' quality of life, promote inclusion and equity and enhance sustainability. Ewing and Cervero (2010) argue that while many cities and states are devoting themselves to spatial planning and urban planning to help reduce car use and related social and environmental costs, the effects of such strategies on travel demand have not been generalized in recent years by the multitude of available studies. Actually, information gathered through Public Participation improves environmental and social planning or results (Koontz and Thomas, 2006). Brown (2015) suggests that by adding the place component, knowledge becomes potentially more usable and influential in planning practice. Meanwhile, digitally supported participation has made enormous strides in recent years. Some excellent reviews critically exist to review a variety of participatory digital platforms or online technologies (Afzalan & Muller, 2018; Falco & Kleinhans, 2018) or participatory apps (Ertiö, 2015). Policies about the transition to sustainable travel are inevitably linked to lifestyle changes. A considerable number of studies have evaluated the factors involved in the choice of modalities of daily travel (such as travel, socio-demographic, built environment factors and, more recently, latent factors) and on the basis of results various urban development strategies and transport policies have emerged with the aim of moving travel towards more sustainable modes of transport. Defining the spatial behavior of individuals is possible through Urban Planning but at the same time, by studying the ways in which people spatially distribute their everyday life, information can be drawn for Urban Planning itself. Therefore, understanding people's mobility is a key to understanding how people live in the city (Adey, Bissell, Hannam, Merriman, and Sheller, 2014).

2   OBJECTIVE OF THE PAPER

The paper aims to explain how the S3URB_ACT[1] Survey experiments with new methods of participation through the use of innovative tools in planning in order to both increase people's experiential knowledge and find sustainable solutions, coherently with the recent directions already stated by the Leipzig Charter on Sustainable European Cities (2007); the UN-Habitat State of the World Cities 2012-2013; and the Habitat III New Urban Agenda (2016). We argue that a planning strategy that is sensitive to the local context and respects the local experiences of the inhabitants can help to find unique solutions and contain conflicts.

The purpose of this study is to test the PPGIS[2] methodology to collect and understand the experiences of citizens, specifically Sapienza students, in order to make the results operational in an integrated planning process in the city of Rome. Attention is paid to the implementation of participatory and experiential input by the interviewees in a useful way in the planning process. This project focuses on sustainable urban mobility that is facing great challenges globally in order to become more sustainable both from an environmental and a social point of view. Increasing the share of less polluting and healthier travel modes, such as cycling, is a top priority in many urban areas. Promoting cycling traffic is, in fact, a means to create a safer, more comfortable and more efficient urban environment. In this way it will be possible to obtain great physical (reducing traffic congestion outside the university city), healthy (contributing to the recommended daily amount of physical activity[3]) and social benefits (becoming part of groups that are sensitive to the topic and to work in the bike workshops currently in progress in different university areas).

This study seeks to improve understanding of the complicated relationships between urban form factors, attitudes and preferences in relation to the specific characteristics of each neighborhood and travel behaviors to examine how the choice of modality differs from the presence of physical constraints or attitudes of people with certain preferences towards specific modes of transport. The study will therefore be of an interdisciplinary nature to improve the planning, design and experimentation of better living environments for citizens.

3   METHODOLOGICAL APPROACH

The proposed methodology of the study places the citizen at the center of this system, as a generator of mobility and user of services, but also as a subject on which impact both positive externalities (accessibility, connectivity, competitiveness) and negative externalities (greenhouse gases, impact local emissions, noise, safety, etc.) produced by mobility. This empirical project proposes the use of innovative instruments of public participation and consultation capable of transferring knowledge derived from research to practice. The data used in this study were collected using the softGIS methodology developed at Aalto University since 2005. SoftGIS is a method of Public Participation in GIS (PPGIS) which combines Internet maps with traditional questionnaires (Brown & Kyttä, 2014). In addition to commonly collected data, such as socio-demographic information, this method allows the collection of location-based data with geographical coordinates, which can be easily viewed and analyzed in ArcGIS/QGIS.

The method is based on online questionnaires that are able to collect quantitative and qualitative geo-referenced inputs. In this way it will be possible to understand how the various groups of students actually use environmental resources in their usual behavior, how much the accessibility of environmental resources is equal and consequently what are the impacts on health. This approach can contribute to a deeper understanding of "people-environment" (PE) research: if user experiences can be assigned to geographic coordinates at a well-defined level, they can also be linked to actual physical settings and therefore to a particular planning

---

1. Sapienza Students Sustainable URBan ACtive Transport
2. Public Participation Geographic Information System
3. Established by the WHO's "Global recommendations on Physical activity for Health"

or design solution. The use, in fact, of the GIS and multicriteria decision analysis (MCA) allows to take into consideration all the perspectives of the interested parties in order to have a greater understanding of the eeds of the citizens and to improve the possibilities of reaching consensus for a more transparent process by giving weight to the final decision.

4 DESCRIPTION

The survey focuses on the choice of home-university travel for ungraduated, graduated and PhD students and researchers (aged 19-40) living in the city of Rome. The survey was designed to gather the socio-demographic characteristics of the respondents, their personal goals based on assessments of the places they visit during a typical week, the modes of transport usually used to reach those places and the frequency of visits. The surveys included both map-based and traditional survey questions. A random sample of students was invited via social internet pages (Facebook, whatsapp groups) and University mailing lists in May 2019. In this survey, respondents used an online interface (Figure 1) to mark on a map everyday home places, study places and activities places. Moreover, the interviewees were asked to answer a series of questions related to the way they perceive their neighborhood[4] and well-being[5].

4.1 Innovative aspects

In this study, a new research strategy will be developed that combines a place-based approach with specific planning solutions and a sensitive study of habitual human behavior. New cartographic visualization strategies will be developed to help planners use place-based knowledge. This research strategy will help to achieve a more realistic and sensitive understanding of environmental health promotion processes, and helps to apply research results in planning practice.

An advantage of online surveys is the possibility of providing multilingual versions, which allow reaching minority language groups that are generally not well represented in traditional public participation processes. The accessibility of respondents is promoted by the usability, visual appearance and scaling of the PPGIS tool from mobile devices to laptops.

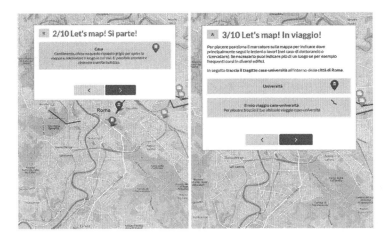

Figure 1. The online interface of the survey where respondents marked on a map their everyday errand points in a typical week.

---

4. From NEWS-A, Neighborhood Environment Walkability Survey-Abbreviated
5. From IPAQ, International Physical Activity Questionnaire

The study of the collected data will allow an analysis and construction of behaviors. It will make it possible to understand how different environments promote health and sustainable behavior and to produce the knowledge that can be used for the development of these urban environments. It will be possible to understand how the various groups of students actually use the environmental resources in their habitual behavior, how much the accessibility of environmental resources is equal and what the consequent health results are. Localized analysis helps define local structural strengths and weaknesses and provides a variety of alternatives for further development. With this new, "soft" information layer available, planners could perform contextually sensitive planning that could fundamentally increase social acceptability of urban densification projects.

To empirically study the hypothesis, students will be represented according to their latent and manifest lifestyle: in addition to the place-based study of students' usual behavior (manifest lifestyle) in their living environment, we will try to study the latent lifestyle dimension, that is, the life models that students can potentially have in another type of context (Heijs et al., 2011).

4.2   *Research and/or technical developments*

With the data collected on the neighborhood it will be possible to study the possible interactions with the walkability index (i.e. the effect of the characteristics of the built environment).

Subsequently, on the basis of a participatory process, it is our intention to plan co-design solutions and effective initiatives that encourage more sustainable behavior based on a comparison of the positive effects on the level of well-being (such as the quantification and mapping of environmental indicators and of the Welfare).

5   RESULT OF THE RESEACH

The survey received approximately 711 geocoded responses, including point data and line data, provided by approximately 195 respondents. Based on the results of the first phase, the new data collection will be carefully planned to ensure that the representativeness of the data (with reference to geographical area, urban structure, population group, lifestyle change) will be as rich as possible.

As previously mentioned, the results of this experimentation will be able to: demonstrate how PPGIS online surveys can be a new tool for participation in the urban planning sector for the production of versatile knowledge (both qualitative and quantitative data; traditional survey data and based on maps, scientific and commented data) and high quality knowledge. Demonstrate how a PPGIS tool can reach a wide range of people; provide additional data and information on travel behavior in order to implement mobility plans; identify co-designed solutions and therefore more accepted and perceived in terms of well-being (quality) by citizens; identify actions to reduce and contain CO_2 emissions and increase mental and physical well-being by improving the urban environment.

6   CONCLUSIONS

The use of PPGIS tools will allow the identification of useful elements to design urban environments that promote quality, health and well-being through the elements identified by the PPGIS tools. The use of tools that mix both physical and digital experiences can represent an innovative approach to improve urban mobility as it can lead to behavioral changes in communities. In fact, it will be possible to understand how the interventions identified on the urban environment, to establish sustainable behaviors and/or attitudes and how they are perceived by citizens in terms of well-being through an posteriori analysis. Indeed, there is evidence that digital tools attract different groups of participants than more traditional tools (McLain, Banis, Todd, & Cerveny, 2017). For this reason, digital tools like PPGIS should be seen as complementary by not replacing the traditional set of participation tools by offering faster and more solid ways to create a channel among the various actors.

As discussed in this document, the results of this study may have important implications for urban planning. For the purpose of a pragmatic approach, the empirical results of this study can be useful to researchers and professionals active in the fields of urban planning, transport and interventions to promote sustainable travel behavior.

However, the greatest contribution of this study is methodological: in fact, by enhancing knowledge research in planning with the introduction of digital maps, it is possible to implement the process of public participation from the early stages of planning. Through the research of the methods and tools used in the development of new services and environments, there is the possibility of improving the quality of life and creating a more sustainable society.

Furthermore, this focus on technology is also proposed as an opportunity to reconcile the difference in knowledge of land use planners and transport planners. By generating a common knowledge base collectively, sectoral differences can be leveled.

The potential application of the results obtained will contribute to an implementation of methods and tools in participatory processes for urban mobility strategies (such as in the SUMPs[6]) in order to develop better user-friendly environments, maximizing distribution of economic resources in the interventions.

6.1 *Barriers and drivers*

This study, although not without limits, can highlight a number of future research directions (active living, daily mobility, planning theory); for example, future studies may be carried out in other contexts and may involve other age groups to understand how the built environment influences lifestyle differences and the choice of how to move.

However, technology remains an obstacle for some users as well as the monetary and human resources required. Presumably this explains low levels of adoption in smaller cities. Although online technologies require less resources, high-quality participatory processes cannot be created without investment. Analyzing PPGIS datasets can be a real challenge, even for planning practices. be totally avoided, since the data processing in planning projects is done by human interaction. It can be partially avoided by having the data correctly and accurately analyzed by an expert or by opening the data for the public.

REFERENCES

Adey P., David Bissell D., Kevin Hannam K., Peter Merriman P., Mimi Sheller M., (2014). The Routledge Handbook of Mobilities. Routledge.

Afzalan N., Muller B., (2018). Online Participatory Technologies: Opportunities and Challenges for Enriching Participatory Planning. Journal of the American Planning Association. 162–177.

Brown G., (2015). Engaging the Wisdom of Crowds and Public Judgment for Land Use Planning using Public Participation GIS (PPGIS). The Australian Planner. 199–209.

Brown G., Kyttä M.,(2014). Key issues and research priorities for public participation GIS (PPGIS): A synthesis based on empirical research. Applied Geography 46. 122–136.

Ertiö T., (2015). Participatory Apps for Urban Planning—Space for Improvement. Planning Practice & Research. 303–321.

Ewing R., Cervero R., (2010). Travel and the Built Environment. A Meta-Analysis. Journal of the American Planning Association. 256–294.

Falco E., Kleinhans R., (2018). Beyond Information-Sharing. A Typology Of Government Challenges And Requirements For Two-Way Social Media Communication With Citizens. The Electronic Journal of e-Government, 16(1). 18–31.

---

6. Sustainable Urban Mobility Plans

Heijs W., Leussink M., Smeets J., van Deursen A., (2011). Re-searching the labyrinth of life-styles. Journal of Housing and the Built Environment. 411–425.

Koontz T., Thomas C., (2006). What Do We Know and Need to Know About the Environmental Outcomes of Collaborative Management?. Public Administration Review 66 (6 supplement). 111–121.

McLain R., Banis D., Todd A., Cerveny L., (2017). Multiple methods of public engagement: Disaggregating socio-spatial data for environmental planning in western Washington, USA. Journal of Environmental Management 204(Pt 1). 61–74.

Walkability and redevelopment

Pedestrians, Urban Spaces and Health – Tira, Pezzagno & Richiedei (eds)
© 2020 Taylor & Francis Group, London, ISBN 978-0-367-46171-3

The walkability of a city with difficult terrain – evaluation of barriers to use of spatial qualities for users and main guidelines for improvement

I. Mrak
Faculty of Civil Engineering, University of Rijeka, Croatia

ABSTRACT: The walkability of a city is an important factor in promoting health in urban environments. In urban environments with difficult terrain, it is a bigger challenge to make it possible for all social groups to make the best use of their environment. The goal of research is to: identify the paths most likely to be used by people with reduced mobility (including chronic illnesses), parts of greatest interest and landscape quality (parks and heritage areas), the overlapping or lack of it, identifying barriers in space, and possible interventions. The approach will be tested on the case study of city of Rijeka, which is largely a vertical city on difficult terrain. The research continues on the research on accessibility and spatial inclusiveness of cultural spaces in Rijeka presented at the conference of DfA Europe network in Pescara in 2018, and the present project Inclusive Campus for University of Rijeka.

1   INTRODUCTION/STATE OF THE ART

The walkability of a city is an important factor in promoting health in urban environments (Forsyth, 2015, Rafiemanzelat, et al., 2017, Singh, 2016, Cubukcu, 2013, Zuniga-Terann, et al., 2017, Talu, Tola, 2018). In some cases, as in urban environments with difficult terrain, it is a challenge to make it possible for all social groups (especially for people with reduced mobility) to make the best use of their environment. The paper continues the research on accessibility and spatial inclusiveness of cultural spaces (Mrak et al., 2018), and the present project Inclusive Campus for University of Rijeka. It also takes into account the input given by persons with disabilities related to the characteristics of public spaces and is a continuation of the ongoing project on inclusiveness and quality of human environment conducted by the author[1] with the support of University Counselling Centre and its' Office for students with disabilities.

2   OBJECTIVE OF THE PAPER

The objective of research presented in the paper is to identify the paths most likely to be used by people with reduced mobility, parts of greatest cultural and landscape quality, their possible overlapping, barriers in space for use of the quality spaces, and possible intervention. The approach is tested on the city of Rijeka, which is largely a vertical city on difficult terrain.

---

1. Including the mentorship of thesis: Cindrić, 2018

3 METHODOLOGICAL APPROACH AND/OR RESEARCH APPROACH

The 1st phase consisted of GIS analysis of 3D terrain model. The lines of the terrain heights were also extracted from this model and used to identify the areas of the paths with smaller slope. Nine areas of this type were identify. Comparing the areas with the street network, most likely paths were identified. The paths were surveyed on site.

In the 2nd phase, based on the analysis of planning documentation and on-site survey, the areas of the city were identified and evaluated based on the characteristics of the heritage. Simple evaluation model was realised through GIS, on the bases of already existing models (Mrak 2013, 2014). The categories were: Cultural heritage (1 – no, 2 – industrial and residential heritage, 3 – historical areas or presence of important singular elements, 4 – historical areas with important singular elements), Green (1 – no, 2 – yes, 3 – green of historic and cultural value), Views (1 – no, 2 – quality views only inside area, 3 – quality views inside and outside area), and Total heritage quality (a multiply of qualities by category). Four maps were produced: Cultural heritage quality, Quality of green elements, Quality of views, and Total heritage quality.

In the 3rd phase, the on site survey was conducted by foot, taking pictures and dimensions of width of passages. The paths were drawn in the GIS and the accessibility values were assigned to coherent segments for different categories (A: Wheelchair users, B: Persons who use other types of aids, C: Persons who have difficulty walking, D: Persons who don't have difficulty walking, E: Persons with sight impairment and low vision, F: Persons with hearing impairment and low hearing, G: Areas for resting; all ranging 1-4: 1- barrier not possible to pass, 2 – great difficulty or lack of safety, 3 - better but not easy, 4 – movement with ease). The total paths were: major paths: 26.140 m, minor/alternative paths: 2.349 m, and total paths: 28.489 m. In the end, maps for different accessibility were produced and compared to maps of evaluation of cultural heritage.

4 DESCRIPTION

4.1 *Innovative aspects*

Innovative aspects are in the combination of topics of walkability and accessibility evaluation and evaluation of heritage. Innovative is also mapping and understanding of the impact that the terrain and other barriers have on everyday mobility for many people – locals, possible tourists, elderly, people with disabilities, as well as understanding the possible positive or negative impact the near environment can have on the amount of benefit somebody can have from the it (for example in relation of Amartya Sen's capabilities concept).

While for some groups (A, B, D, E) there are legal standards that can be considered for barriers, for group C some aspects were considered: slope, length of slope, places for pausing and rest, but also the congestion of pedestrian traffic (which may cause loss of balance and disorientation) and difficulty to cross the roads in time and with ease (access to road with a step). This opened different and more realistic interpretation of walkability that the one using only basic elements like elements of vertical difference or width of the path.

4.2 *Research and/or technical developments*

Persons with visual impairment and low vision have very different tactics for orientation and movement. In this research their ease of movement is addressed as combined. In different types of environment, the separation of categories could be advisable.

Using GIS several accessibility and walkability maps were produced. The approach allows also for overlap of various thematic maps (more than 40 maps), statistical analysis and quantification of characteristics of paths.

5 RESULT OF THE RESEARCH

On site survey showed that almost no pedestrian paths are wide enough for 2 users of group A to pass each other. The issues regarding groups B and C opened richer and more realistic interpretation of walkability and is certainly a bases for further research. Group D can mostly go in all considered locations, but some of the paths are mixed traffic (with no sidewalks) and therefore producing safety concerns even for the traditionally "standard" user. There are no elements of orientation for group E. For regular users of group E, the environment is accessible only in small amount, but if non-locals are considered, the results would be 0. For group F, the safety is the most important issue. Places of rest are lacking.

The distribution of Total heritage quality is somehow different respect the general idea that the heritage areas of most value are the centre and portual areas. This is largely related to the little presence of green in those areas, and high presence of green and good views in some other areas; as represented in Figure 1.

In comparison with group D, in areas of highest Cultural heritage values, the least access is possible for group A (9%) and group E (22%). Only 3% is accessible to all. In areas of highest Green values, the least access is possible for group A (6%) and group E (16%). Only 5% is accessible to all. In areas of highest View values, the least access is possible for group A (6%) and group E (16%). Only 3% is accessible to all. For other users, the rate is around mostly 40-50%, which is little considering difficulties of movement for many different users – children, elderly, people with travel or shopping gear... This is especially so when considered that the choice of paths to analyse was based on the lower slope than usual for the city of Rijeka, to begin with. Not all paths are easily accessible even for group D mostly due to paths of mixed traffic and safety (91%). Similar scores are obtained also considering only areas of highest heritage values. For most groups, the most accessible are areas of high cultural values and less of green and views. The accessibility varies the least for groups A, E and F, such as in Table 1.

Considering the diversity of barriers and different groups' needs, and big number of coherent segments, the guidelines for improving the access are more of the procedural than technical. The priority of interventions should be based on sequent criteria: heritage value of the area and already possible (higher value) of accessibility for different groups. This overlap makes area of Pećine, center and area near Delta priority for intervention. All interventions should be done with the process of public participation during planning and design process, and adapt in time. The interventions have to consider complex approach (public transport, barriers for different groups...) and complete traffic flow. From the analysis it can be seen that a different concept of traffic could sometimes be necessary to allow for wider paths and more places for rest that would allow for passing by and ease of use for many groups of users that today can hardly benefit from the existing paths.

Figure 1. Paths evaluation for persons who have difficulties walking (worst:red-best:green), on the map of total heritage value (lowest:pale-highest:darkest color).

Table 1. Highest quality areas by accessibility.

	Access. for group	Partially acces.	Not acces.
West.area	C, D, F	A, B, E	
Cent. area	B, C, D	A, E, F	
Port	A, B, C, D, F	E	
River area		C, D	A, B, E, F
Trsat	B, C, D	A, F	E
Bulevard	Not considered due to slope		
Pećine	B, D	A, C, E, F	

6 CONCLUSIONS

The evaluation and mapping of the characteristics of the terrain was possible mostly due to the previous in depth research about types of barriers for different users. The ongoing communication and long term activities would help with improving the knowledge of barriers and opportunities presented by the environment.

Further development could be in: continuous platform for registering the barriers and opportunities, positive and negative aspects, introduction of ITC for input of barriers.

The research would be certainly useful also in other contexts, especially with high presence of cultural heritage and difficult terrain.

6.1 *Barriers and drivers*

The quality of 3d terrain data is generally useful for basic slope detection, but not enough for identifying barriers in space. Other important barriers to the approach are multiple types of spatial/physical barriers for different groups which necessitate understanding of movement and barriers. Especially challenging are dynamic barriers (such as parked cars) and very short parts of coherent characteristics.

Drivers to develop similar approach are the rise of ICT that could allow for more detail mapping and communication. At this moment, services like Google Earth are still not dynamic and detailed enough to be used as primary source of data. Therefor there is a necessity to work on other types of data gathering. The mapping technique could also enrich the model with mapping the positive characteristics of the paths (like shades), or open the possibility of citizens' proposals for placing elements of spatial improvement. Developing such platform could be the basis for virtual guide and as input to public administration about the real state of barriers.

REFERENCES

Cindrić, M. (2018), Univerzalni dizajn u arhitekturi za prostore društvene namjene. Bachelor thesis, University of Rijeka, Faculty of civil engineering.

Cubukcu, E. (2013). Walking for Sustainable Living. Procedia - Social and Behavioral Sciences, Volume 85, 2013, 33–42.

Forsyth, A. (2015). What is a walkable place? The walkability debate in urban design. Urban Design International 20, no.4: 274–292.

Mrak, I., "An Evaluation Model For Cultural Heritage In Spatial Planning", Inderscience International Journal of Global Environmental Issues (IJGEnvI) special issue "Cities as Sustainable Wealth Creators", 2014, 13 2/3/4; 206–234.

Mrak, I., "A Methodological Framework Based on the Dynamic-Evolutionary View of Heritage", MDPI's Sustainability 2013, 5(9), 3992–4023, <http://www.mdpi.com/2071-1050/5/9/3992>.

Mrak, I., Nuždić, S., Franković, M., Matan, C. (2018) "Preliminary research: Inclusiveness of cultural heritage - a model of evaluation", Cultural Heritage for All Unlocking Europe's potential through Design for All, Pescara (Italy), 28–29.11.2018, poster.

Rafiemanzelat, R., Imani Emadi, M., Jalal Kamali, A., (2017). City sustainability: the influence of walkability on built environments. Transportation Research Procedia, Volume 24, 2017, 97–104.

Singh, R. (2016). Factors affecting walkability of neighborhoods, Procedia - Social and Behavioral Sciences 216 (2016) 643–654.

Talu, V., Tola, G. (2018). La citta' per immagini. Verso la definizione di un insieme di requisiti spaziali per la progettazione di citta' autism friendly. LIST LAB Trento.

Zuniga-Terann, A. A., Orr, B. J., Gimblett, R. H., Chalfoun, N. V., Marsh, S. E., Guertin, D. P., Going, S. B. (2017). Designing healthy communities: Testing the walkability model. Frontiers of Architectural Research, 2017 6 63–7.

Pedestrians, Urban Spaces and Health – Tira, Pezzagno & Richiedei (eds)
© 2020 Taylor & Francis Group, London, ISBN 978-0-367-46171-3

Spillover effect of urban regeneration on pedestrian accessibility and walkability

M. Tiboni & F. Botticini
Department of Civil Engineering, Architecture, Land, Environment and Mathematics (DICATAM),
University of Brescia, Italy

ABSTRACT: The paper wants to explain some of the first results of a work that aims to show which are the urban transformations effects on pedestrian accessibility and walkability. Particularly, the work aims in defining a methodology suitable to measure the spillover effect of urban regeneration on urban spaces and city users.

According to the 11th Sustainable Development Goal, given by United Nations with the Agenda 2030 for the Sustainable Development, who asks cities and human settlements to become safe, inclusive resilient and sustainable, the methodology aims to give a tool that can help in monitoring if changes of the structural shape of cities are going into the direction drew by the Agenda.

The developed methodology is subdivided into three steps: quantification, assessment and monitoring. The first one starts with the deliberations analysis from which it was possible to define the process behind every urban transformation. In the second step each transformation has been mapped, with a linked database, containing the data recorded, such as the intervention type, the cost or the year in which the building site started. The third phase is the monitoring and it let to assess the effects of these transformations in time.

The case study of the work is the process of diffuse urban regeneration that occurred in Brescia between 2013 and 2018. In this period plenty of urban operations have been done. Some of these operations are still happening now and they are deeply changing districts shape.

Attention was given to the analysis of pedestrian and cycle paths. Particularly, it is important to observe how they contribute in creating a network of clean mobility inside and between districts promoting and enhancing a sustainable and inclusive lifestyle, as it is suggested by the United Nations.

1 INTRODUCTION: OPEN SPACES AS KEY FOR SUSTAINABILITY

The most part of reports about urban development agree in saying that up to 2050 about 80% of people will live in cities (ASVIS, 2019), (EU, 2016). This shows that the challenges related to sustainable development must take on in urban areas that have to change their structure to face new external efforts such as the one related to climate change or to social inclusion (ONU, 2015). Namely, as the One Plane City Challenge report by WWF says: *"Cities have the richest opportunities to accelerate positive change by their planning of spatial structure, connected infrastructure, and organizational and social dynamics, and by the possibility for scale in effecting solutions"* (WWF, 2017). That means that there is a strong link between shape, infrastructures and quality of life inside the urban areas (Codispoti, 2018). Especially, cities can directly improve quality of life reducing environmental impacts (WWF, 2017) while mobility and accessibility are key areas for delivering solutions for a climate-resilient future, built on 100 percent renewable energy, and for the creation of attractive, sustainable cities, based on health, equality, and improved life quality (WWF, 2017). These researches highlight that to foster the sustainable goals given by the Agenda 2030, particularly the ones related to urban development and climate change adaptation (ONU, 2015), it is important to operate on the physical matrix of cities creating open public spaces that could be useful and nice at the same time (Yaro, 2009).

2 OBJECTIVE OF THE PAPER

Sustainability is composed by three dimensions: environmental, social and economic (ONU, 2015) and interventions of urban regneration help in fostering all these aspects. The paper wants to analyse which is the contribution of a process of urban regeneration in achieving the sustainability goals. Urban infrastructure works can, at the same time develop a network of green and clean mobility and create open spaces in which it is possible to meet other people fostering social inclusion too (Tiboni & Botticini, 2018). A process of urban regeneration can contribute in two ways in increasing accessibility: by strengthening the existing infrastructures, creating a continuous path of clean mobility and by fostering the diffusion of services and reducing the distances between commercial activities and residential areas. Both the solutions contribute in creating urban spaces which are safer for pedestrian and other weak users of street. The article focuses on the urban shape of the city and how that shape is changing after that a process of urban regeneration has started. A second aspect this article wants to focus on is the relationship between public interventions on open spaces and the willingness of private owner and stakeholders to participate in the process fostering the third dimension of sustainability: the economic one. It is well demonstrated in literature that increasing the accessibility of an area through the creation of infrastructures, such as Public Transport Lines, contributes in increasing the willingness of private owner to invest in the development of the site (Medda, 2012), (Munoz-Gielen & Tuna, 2010) and this aspect has a huge influence on the value of the land and on real estate market (Auzinis & Viestrus, 2017)

3 METHODOLOGICAL APPROACH

Transformations of districts shape find their origins in private interventions which can be sub-divided in two categories; the first type is called big operations and concerns real estate or commercial interventions. This actions entail urban infrastructure works such as streets redevelopment with the realisation of cycle or pedestrian paths, the creation of new car parks or the refurbishment of green areas that have been transformed into parks for families and residents. There is a second type of transformations in which there aren't any urban infrastructure works related to private operations. These small interventions concern changing in the use of buildings and contribute in creating new services or commercial activities inside districts (Tiboni & Botticini, 2018).

To define what is linked to every transformation it was necessary to start from the deliberations analysis. Thanks to the use of a GIS software it was feasible to map data contained in the deliberations, concerning the cost of each project, the typology of intervention and the year in which the works started. In the attribute table of the GIS sotware it was feasible to create a univocal code, called ID code (Errore. L'origine riferimento non è stata trovata.), that allows to link each project with the deliberations and define which was the administrative iter behind every urban operation (Tiboni & Botticini, 2018).

The creation of this database allows to do some analysis about how districts shape is changed across the examined five years. The output of this work is a matrix that can be called "wealth in urban environment index", which allows to define life quality in urban areas is composed by two parts: in ordinates there are the three dimensions of sustainability while in absissas there are the aspects that characterize urban environment. These aspects are subdivided into five area of analysis: built environment, real estate, society, transportations and policies. Each area is subdivided into indicators and each indicator is composed by criteria. The last level is composed by parameters which are necessary to assess the criteria and the indicatoros. Each parameter is an action that the urban planner or the local body can do to achieve the goal of sustainable development (Errore. L'origine riferimento non è stata trovata.).

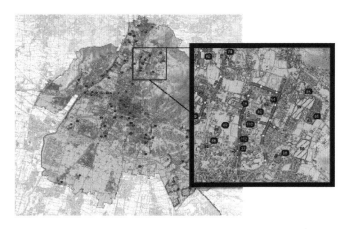

Figure 1.   Urban infrastructure works linked to private interventions.

Ai [fields of analysis]	Iij [indicators]	Cijn [criteria]	Pijnm [parameters]	S [dimensions of sustainability]		
				S1	S2	S3
	Ii1	Ci11	Pi111			
			Pi11m			
		Ci1n	Pi1n1			
			Pi1nm			
	Iij	Cij1	Pij11			
			Pij1n			
		Cijn [criteria]	Pijn1			
			Pijnm [parameters]			

Table I.   The "wealth in urban environment matrix".

Obviously there are two possible way to read the matrix: the first is horizontal and with this method it is possible to obtain how a single parameter (or action) contribute in stenghtening the aspects of sustainability. The second way is the vertical one and if one dimension of sustainability is fixed, it is possible to see which are the actions that maximise that aspect.

By applying the matrix to the scenario 0 related to the ex ante situation (before the urban transformations) and to the scenario 1 related to the ex post situation (after that the process has started) it was possible to obtain a qualitative measure of how the situation has changed.

The possible developments of the methodology here described concern in evaluating how different projects can interact each other and obtain a spillover effect on the city. In other words it is necessary to define which are the outcomes of interventions done in different areas on the entire territory analysed and how they contribute in creating a network. The other aspect that it is interesting to study is how the incrised quality of life in urban areas due to a project or a group of transformations can be quantified and monetised. This is an important aspect in the optic of economic sustainability because it allows to define a land value of private areas that can be captured by public bodies and redistributed and reused to finance other urban regeneration processes.

4 FIRST RESULTS OF THE RESEARCH

The case study chosen is the urban regeneration process happened in Brescia between 2013 and 2018. In these five years there was a changing in the local public bodies' development strategies. This new green vision aimed to foster the sustainable development of the land reducing free soil consumption and enhancing the areas that had been already urbanised (Tiboni, 2015). Thanks to a deep collaboration with stakeholders plenty of transformations started in that years and they are still changing districts shape. The analysed operations are the ones taken in exam by the Urban Planning Sector of Brescia Municipality during the defined period.

The first results of this research show that urban transformations and, particularly, urban infrastructure works linked to private investments can transform districts creating a green mobility networks (an example in Figure 2 and Errore. L'origine riferimento non è stata trovata.). These networks can contribute in pursuing the sustainability goals through deep changing in the physical matrix of cities. The new cycle and pedestrian paths link existing networks with inner districts areas in which there are services or commercial activites and they contribute in creating better places in which residents can move, live and meet other people.

Figure 2.   Via Corsica before the regeneration of urban space (2017).

Figure 3.   Via Corsica after the regeneration of urban space (2018).

5 CONCLUSIONS

This article aims to explain the first thoughts about how a municipality can develop clean mobility infrastructures taking into account aspects regarding sustainability and develop an urban regeneration process that is widespread around the city and that can have a huge link with the real estate market. This process can foster at the same time all the three dimensions of sustainability: creating a green mobility paths can help in solving some prolems related to environmental aspects such as noise, pollution, greenhouse gasses emissions and other aspects related to climate change. This clean network contributes in creating better places for citisens and residents and this aspect foster social inclusion too. The other topic which is necessary to focus on is that deliberations contirbute in simplifying the realisation of new services or commercial activities inside districts and this is another aspect that can help in creating places in which people can meet. The last aspect regards topics related to economic sustainability. This urban regeneration process is developed with partnerships between private stakeholders and the local administration, increasing redevelopment of public open spaces and life quality, which can be a further stimulus for private owners.

REFERENCES

ASVIS. (2019). città e comunità sostenibili. Tratto da asvis.it/goal11.
Auzinis, A., & Viestrus, J. (2017, Novembre). A values-led planning approach for sustainable land use and development, November. Baltic journal of real estate economics and construction management, 275–286.
Busi, R. (2009). Per una città più sicura. Ed amica. E più bella. TEMA.
Caprì, S., Ignaccolo, M., Inturri, G., & Le Pira, M. (2016). Miglioramento dell'accessibilità pedonale attraverso "green walking networks" per contrastare il cambiamento climatico. Green walking networks for climate change adaptation. Transportation research part D: transport and environment, 45, 84–95.
Carmona, M., & Siehl, L. (2004). Measuring quality in planning. Managing the performance process. Oxon Spoon Press.
Carmona, M., Heath, T., Oc, T., & Tiesdell, S. (2003). Public Places, Urban Spaces. Oxford: Architectural Press.
Codispoti, O. (2018). Forma urbana e sostenibilità. LISTLAB.
Congiu, T., & Plaisant, A. (2018). The role of connective space in regeneration. Urban Design, 18–20.
EU. (2016). Urban Agenda for the EU; Pact of Amsterdam. Amsterdam.
London Municipality. (2017). Land Value Capture. Final Report. London: Tranport for London.
Medda, F. (2012). Land value capture finance for transport accessibility: a review. Journal of Transport Geography, 154–161.
Munoz-Gielen, D., & Tuna, T. (2010). Flexibility in planning and the consequences for public values capturing in UK, Spain and the Netherlands. European Planning Studies, 1097–1131.
O'Flynn, J. (2007). From New Public Management to Public Value:. The Australian Journal of Public Administration,vol. 66, no3, 353–366.
ONU. (2015). Transforming our world: the 2030 Agenda for sustainable development.
Tiboni, M. (2015, Febbraio). Verso un nuovo PGT per la città di Brescia. Architettura e Paesaggio, n1.
Tiboni, M., & Botticini, F. (2018). Gli effetti delle previsioni urbanistiche sulla rigenerazione urbana diffusa. Il caso di Brescia. XXXIX Conferenza Italiana di Scienze Regionali. Bolzano.
Tira, M. (1997). Pianificare la città sicura. Roma: Edizioni Librerie Dedalo.
Tira, M. (2001). La localizzazione degli attraversamenti pedonali. In R. Busi, & L. Zavanella, Tecniche per la sicurezza in ambito urbano. La protezione del pedone negli attraversamenti pedonali (p. 93–101). Forlì: EGAF.
WWF. (2017). Urban solutions handbook. Gland: WWF International.
Yaro, R. (2009). A plan for alla reasons. Regional Plan Association.

Pedestrians, Urban Spaces and Health – Tira, Pezzagno & Richiedei (eds)
© 2020 Taylor & Francis Group, London, ISBN 978-0-367-46171-3

Pedestrian mobility as urban regeneration strategy

E. Conticelli, E. Bruni & S. Tondelli
Department of Architecture, Alma Mater Studiorum - University of Bologna, Italy

ABSTRACT: This paper analyses planning strategies and design solutions based on the promotion of walkability for regenerating the urban environment. Starting from the analysis of different virtuous European case studies, where urban regeneration policies have been linked with the promotion of soft mobility, some key factors affecting walkability have been detected: livability, safety, attractiveness and accessibility. The proposed analysis has been carried out both at urban and local scale, by considering a set of indicators easy to be measured that enabled to describe and measure the four main identified factors, thus providing insights for activating similar policies in other contexts.

1 INTRODUCTION

In the last decades, greater attention has been dedicated by city planners and policy makers to the effects and possible risks of car-oriented urban development models, facing the urgency of supporting measures aimed at restoring public spaces for citizens, through the creation of a more dynamic, livable and economically attractive urban environment (Sastre et al., 2013). Starting from considering the most acknowledged benefits in terms of regeneration of urban places achieved by the promotion of walkability, the paper investigates valuable case studies of walkability promotion in order to identify walking-oriented strategies and design solutions to increase the quality of the urban environment. Measures aimed at encouraging greater pedestrianization in highly urbanized contexts may have a great regeneration potential of the built environment, achieving to beneficial results in terms of increased livability (Harvey and Aultman-Hall, 2016), safety and security (Southworth, 2005; Ruiz-Padillo et al., 2018), attractiveness (PQN, 2010; Taleai and Yameqani, 2017) and accessibility (Ewing and Cervero, 2010) of public spaces.

Livability is a condition characterized by high quality of life with low levels of noise and air pollution and one key ingredient is to create a walkable environment (Shamsuddinet al, 2012).

Safety and security. Safety is a condition strongly related to urban traffic, while security is associated with the perception and risk of crime in a certain environment (Foster et al., 2013). By giving priority to pedestrians and increasing activities in public spaces, both road safety and security can be consequently improved.

Attractiveness is related to the quality of the urban environment and the presence of amenities, that can be easily perceived and used by pedestrians, increasing the sense of belonging to a place.

Accessibility: is the ability to reach desired destinations, making an acceptable effort (Erath et al., 2017; Habibian and Hosseinzadeh, 2018). Investing in pedestrian mobility ensures a greater accessibility for all.

The above mentioned factors have a twofold role: they are crucial for ensuring the high quality of the urban environment and strongly affect the decision to walk. To better understand the components of these four key factors, a case study analysis has been carried out and specific indicators have been considered.

2 SUCCESSFUL PEDESTRIANIZATION EXPERIENCES IN EUROPE: URBAN STRATEGIES AND DESIGN SOLUTIONS

The proposed case study analysis has been conducted on two different scales, to highlight general strategies and design solutions.

At urban scale, four different case studies (Copenhagen, Nuremberg, Strasbourg, Wolverhampton) have been analyzed where walkability has been extensively improved and promoted, representing successful experiences with positive benefits at urban scale. At this scale, *accessibility* represents the most relevant and mesurable factor that therefore has been assessed through specific indicators.

At local scale, three other case studies (Brighton, Velenje, Heidelberg) have been considered, where specific and succesful interventions on existing streets have been undertaken, leading to a general improvement of walking and urban conditions. Other indicators have been selected for detecting the most common features in a pedestrian environment, influencing *accessibility, attractiveness, livability* and *security*.

2.1 *Successful strategies at urban level*

The selected case studies are widely recognized as good example of succesfull pedestrianization through the enhancement of accessibility by ensuring good public transport connections as well as parking areas in the surroundings.

Copenhagen (DK). One of the first European cities to pursue strategies aimed at reducing private vehicular traffic to make the historic center accessible to soft mobility. The strategy began in the 1960s and today the pedestrian area in the city center is around 100,000 sqm.

Nuremberg (DE). Between the 1960s and the 1970s the city carried out strategic actions to drastically reduce air pollution caused by car traffic, based on the progressive pedestrianization of the historic center and the construction of a metro line. In 1988, the closure of some important central roads led to a decrease of traffic of 71% in the city centre, improving air and living conditions.

Strasbourg (FR). Since the 1990s the administration decided to apply a strategy to drastically reduce the presence of cars within the historic center. Firstly, two tramways which served the center had been built accompained by the gradual closure of the central roads to private vehicles. A wide awareness campaign was launched as well to encourage citizens and traders to accept the changes. A monitoring process accompanied the implementation of the pedestrianization plan registering its success.

Wolverhampton (UK). Between 1987 and 1991 a strategy strengthening public transport was implemented, making the city centre accessible only by buses, taxis, pedestrians and cyclists, with some exceptions for service vehicles. New parking lots were opened in the center as well. These measures were accompanied by a monitoring programme, showing important results.

The improvement in accessibility in the four case studies have been assesed by considering specific indicators commonly used to describe this factor (Ewing and Cervero, 2010):

- *Percentage of pedestrian area:* this indicator measures how much space is exclusively destined to pedestrians compared to the entire area of the city center.
- *Density of pedestrian network:* representing how pedestrian network is branched into the urban grid.
- *Density of breaking points:* ratio between number of intersections with vehicular roads and its length. The more this parameter is high the more the pedestrian network is perceived as less accessible.
- *Connectivity:* ratio between number of intersections and length of the pedestrian network, representing how pedestrian network is branched and capillary.
- *Public transport stops density:* ratio between the number of PT stops and the extension of the pedestrian area.

- *Density of parking spaces:* ratio between the number of parking spaces and the extension of the pedestrian area.

Strasbourg city center has shown the best performances: it has a very high connectivity and fewer points of discontinuity, ensuring a homogeneous, widespread and well connected pedestrian network.

Densities of parking lots and public transport stops highly encourage a modal interchange towards walking.

Wolverhampton has a less developed pedestrian area, nevertheless it shows high accessibility in terms of density of PT stops and availability of parking spaces, which strongly supports the business actvities settled in the city center.

Copenhagen and Nuremberg have pedestrian networks with higher densities of breaking points than the other two city centres. Nevertheless, they have an excellent accessibility in terms of density and spatial distribution of public transport stops and parking lots, showing an attitude for a further extension of the pedestrina network.

2.2   Successful strategies at local level

When assessing the walkability of a place, it's important to consider the general quality of the urban environment, that is made by different elements. Thus, the second part of the analysis focoused on considering most common pedestrian facilities and conditions that affect the decision to walk. Three successful case studies of urban environments recently redesigned have been analyzed at local level to identify solutions and key elements that increased attractiveness, livability, security and accessibility.

New Road, Brighton (UK). New Road was traditionally an important connection between the first Brighton's outskirts with the city centre. The City Council decided to give priority to pedestrians and consequently to redesign the street to reduce unsafety and insecurity conditions. The new solution was a pedestrian area where cars and pedestrian share the street space. Consequently, car traffic drastically decreased, and pedestrians and cyclists increased as well as commercial activities.

Promenada, Velenje (SI). This street closed to vehicles during the 80s, but a real renewal has not succeeded, letting the street with no identity and social activities. In recent years the entire area has been redesigned, transforming the street into an urban park, with many micro-spaces for resting

Bismarkplatz, Heidelberg (DE). The square has always been representing a crucial multi-modal hub of the city but characterized by many security problems due to unsafe pedestrian conditions. Vehicles have been dominating the square, with limited sidewalks and platforms for commuters waiting for public transports. Thus, the square was completely redesigned, creating a "pedestrian island" in the centre of the square, equipped with trees, benches and dedicated lanes for trams and buses. Underneath the square a multi-level car park has been built to ensure multimodality and accessibility.

The selected case studies are characterized by very different environments, but there are common features and facilities that can be highlighted and measured. These elements can be considered as minimum requirements for ensuring good livability, attractiveness, secuity and safety and accessibility at local level. Also in this analysis, specific and easily measurable indicators have been selected:

- *Density of amenities and commercial activities.* Number of activities in the study area, indicating the level of attractiveness of a place due to the presence of restaurants, bars and shops. Security and livability are affected by the presence of these activities as well.
- *Tree shadow coverage.* Presence of tree in the study area, offering sunlight protection during summertime and adorning the urban environment, thus improving livability and attractiveness.
- *Presence of street furniture.* Number of appropriate urban features (e.g. bins and benches) in the study area. The presence of these elements influence attractiveness, inviting people to stop for resting and talking.

- *Presence of bicycles racks*. Number of racks in the study area, encouraging exchange between soft modes of transport, positively influencing accessibility safety and attractiveness.
- *Presence of public lightings*. Street lighting coverage in the study area is a facility that drastically improve security and safety as well as attractiveness and livability of a place making it more vibrant also during the night.

The above mentioned indicators have been measured in the three case studies gathering very different values. Although all cases have urban furnitures, commercial activities, trees, street lightings etc. the intensity of their presence is different because these places have different urban nature and purposes. *New Road* is characterized by a high presence of shops and restaurants where people can easily rest and seat, therefore it has high levels of amenities and furnitures. *Bismarkplatz* was designed for being an efficient and pleasant transport node; indeed it shows higher value of density of public lighting, that is necessary for ensuring that the transport hub is safe and secure especially during the night, and tree coverage, increasing the beautyness of the square. *Promenada* has shown good values in tree coverage and presence of furnitures and public lighting due to its nature of pedestrian path and urban park.

3 CONCLUSIONS

By considering the four main factors that are currently considered as more relevant in affecting the decision to walk, the research has carried out a case study analysis to extract possible solutions ensuring more accessible, livable, secure and attractive urban spaces, thus encouraging regenerative processes.

The case study analysis has shown similar approaches in terms of pedestrianization strategies and design solutions and has allowed to identify some key elements to be considered for future interventions.

In order to promote a more sustainable mobility and accessible urban spaces, a general reccommendation is to improve and encourage the use of sustainable means of transport, improving and expanding a multimodal transport system. This principle is addressed by ensuring an effective exchange between foot and other means of transport, by encouraging the built of suburban parking areas and/or central underground parking areas, the good positioning and equipment of the interchange nodes, and by promoting a general improvement of pedestrian paths.

The most promising measures for increasing livability and attractiveness of an urban area are pedestrianizations, accompanied by awareness campaigns and based on gradual interventions. Different intensity of pedestrainizations can be proposed: from full pedestrianisations, to areas for pedestrians and public transport, to shared spaces till minor changes to improve pedestrian facilities.

Other key solutions aim to increase safety and security. They mainly concern the promotion of higher pedestrian crossings visibility, pedestrian crossing narrowing, pedestrian islands, higher quality of public lighting. These solutions are mainly targeted to improve the interactions between pedestrians and other motorized means of transport but also to reduce unsocial behaviors.

The use of indicators has been useful to compare the different case studies trying to highlight common rules and differences. Apparently there are no quantitative rules in terms of targets to achieve for planning and designing pedestrian spaces but rather the identification of general reccomendations and solutions that could support public administrations and designers to plan and design walkable and high quality urban spaces.

REFERENCES

Erath, A., Eggermond, M., Ordóñez, S., Axhausen, K. 2017. Introducing the pedestrian accessibility tool: Walkability analysis for a geographic information system. Transportation Research Record: Journal of the Transportation Research Board, 2661, 51–61.

Ewing R., Cervero R. 2010. "Travel and the Built Environment", in Journal of the American Planning Association, no. 76(3), pp. 265–294.

Foster, S., Wood, L., Christian, H., Knuiman, M., Giles-Corti, B. 2013. Planning safer suburbs: Do changes in the built environment influence residents' perceptions of crime risk? Social Science & Medicine, 97, 87–94.

Habibian, M., Hosseinzadeh, A. 2018. Walkability index across trip purposes. Sustainable Cities and Society, 42, 216–225.

Harvey C. & Aultman-Hall L. 2016. Measuring Urban Streetscapes for Livability: A Review of Approaches, The Professional Geographer, 68: 1, 149–158.

PQN - Pedestrian Quality Needs. 2010. Final Report of the COST Project 358, Walk21, Cheltenham. Available at: www.walkeurope.org (September, 2018).

Ruiz-Padillo, A., Pasqual, F., Uriarte, A., Cybis, H. 2018. Application of multi-criteria decision analysis methods for assessing walkability: A case study in Porto Alegre, Brazil. Transportation Research Part D, 63, 855–871.

Sastre J., Sastre A., Gamo A., Gaztelu T. 2013. "Economic Impact of Pedestrianisation in Historic Urban Centre, the Valdemoro Case – Study (Spain)", in Procedia – Social and Behavioral Sciences, no. 104, pp. 737–745.

Shamsuddin, Shuhana, Rasyiqah Abu, and Siti Fatimah Ilani. 2012. "Walkable Environment in Increasing the Liveability of a City." 50(July): 167–178.

Southworth, M. 2005. Designing the walkable city. Journal of Urban Planning and Development, 131(4), 246–257.

Taleai, M., Yameqani, A. 2017. Integration of GIS, remote sensing and multi-criteria evaluation tools in the search for healthy walking paths. KSCE Journal of Civil Engineering, volume 22, issue 1, pp. 279–291.

Pedestrians, Urban Spaces and Health – Tira, Pezzagno & Richiedei (eds)
© 2020 Taylor & Francis Group, London, ISBN 978-0-367-46171-3

Urban accessibility as an approach for the regeneration of urban peripheries. The experience of the Sant'Avendrace district in Cagliari

T. Congiu, V. Fais & A. Plaisant
Department of Architecture, Design and Urban Planning, The University of Sassari (Sardinia), Italy

ABSTRACT: Pheripheral areas are conventionally depicted by physical and social degradation, poor services, lack of public spaces.

In the project of urban regeneration for Sant'Avendrace district in Cagliari (Italy), accessibility played a key role in connecting functions and spaces, thus reviving the marginal character of the area.

The reported experience describes the operating framework of interventions proposed as a mean of implementation of a national urban Program, paradigmatic of an important change in the approach to urban regeneration policies.

1 STATE OF THE ART

In Italy, since the 1950s, the succession of different urban regeneration programs resulted in a multiplicity of fragmented solutions because of the heterogeneity of the managing authorities, combined with speculative drives for the low cost of land (Salzano, 2000). There have been several attempts to restore marginal situations, such as the '90s "complex" urban programs and various funding programs of the 2000s.

On the contrary, the multifaceted concept of periphery involves diverse dimensions. It refers to a state of physical, social, environmental and cultural decay (Belli, 2006), which result hardly to be described by mean of synthetic indices and interpretative categories according to spatial organization models[1].

Starting from the awareness of the limits of the extraordinary programs, which channelled large resources to public works, the focus is on the supply and the connection of new urban functions, uses and services in terms of opportunities for the growth of urban quality.

2 OBJECTIVE OF THE PAPER

This paper describes the operational framework of interventions for the Sant'Avendrace district as a mean of implementation of the Italian 2016 Extraordinary Program of intervention[2]. Planning choices in the short and long term were informed by a series of strategic objectives tailored to the context: 1. Rethinking urban accessibility at different scales; 2. Strengthening urban resilience capacity; 3. Improving urban space liveability; 4. Promoting sustainable lifestyles and behaviours.

---

1. In the "City Plan" (DPCM 26/10/2015) an area of intervention is outlined by the index of social unease and the index of building degradation, taking into account the average of deviations from the national average values of employment, unemployment, youth concentration, schooling rates and the state of conservation of the degraded urban area buildings.
2. Presidenza del Consiglio dei Ministri, Bando per la riqualificazione urbana e la sicurezza delle periferie, DPCM 25 maggio 2016.

Operationally, a system of physical and functional connections was designed to restore the relationships between the suburb and the dominant elements of the urban and environmental territorial system.

3   RESEARCH APPROACH

An operational framework (Figure 1) was developed to support planners and designers in the definition of specific interventions: the 5 general types of actions provided by the 2016 Program have been articulated into operational objectives and context-oriented "project attentions", thus avoiding the fragmentary nature and discontinuity of interventions.

More precisely, an in-depth analysis of the study context and a review of existing local planning tools was accomplished with the aim of identifying the key aspects of urban organisation (the "project coordinates"). In order to make the strategy effective and synergistic, an integrated Program of interventions was provided with the aim of planning the interventions along time, according to their ability to activate changes, integrate and strengthen their initial effects or complement them (*activating, supporting and complementary projects*). An evaluation and monitoring activity by mean of performance indicators was envisaged to measure advancements and verify the effectiveness of interventions in the long term (Figure 2).

The attention to the process and its multi-scale interconnections, as well as to the continuity and coordination of interventions over time, represents one of the most innovative aspects even recognised for its awards[3].

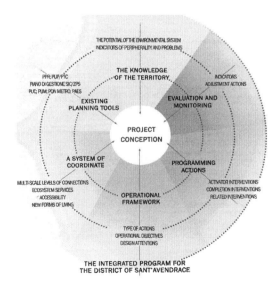

Figure 1.   The methodological process.

3. The proposal for Sant'Avendrace district ranked 23/120, was awarded and funded with 17.995.170,00 €. The project was also positively evaluated by the ANCI for the experimental contents in terms of project activation (ANCI/Urban@it dossier).

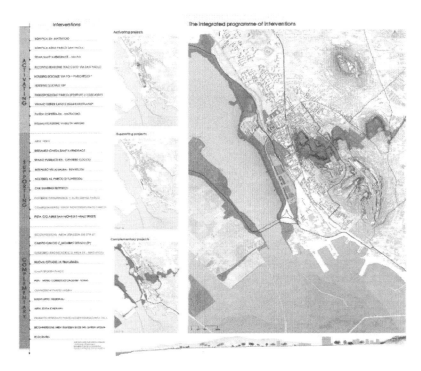

Figure 2. Planning phases and future scenarios for the metropolitan area of Cagliari.

4   THE PROJECT PROPOSAL

The landscape matrix and the spaces for mobility assume a strong potential in establishing relationships of different nature and levels.

From this basis, a renewed system of physical, functional and perceptive connections was developed, making different parts of the city accessible and strengthening the relations between the neighbourhood and the new urban functions, services and public spaces (e.g. sporting and educational park, social housing complex). A new branched system of public spaces and facilities, that enriches urban and ecosystem services supply and promotes collect-ive uses and sustainable lifestyles, is obtained by redesigning the roads network and its sur-roundings, including interstitial abandoned areas (Congiu *et al.*, 2018).

The proposed system of connections hinges on the redevelopment of the main district axis and its transversal viability. In particular: 1. keeps its role as an arterial road which handles flows at the urban and metropolitan scale; 2. gives back to the axis the original character of urban avenue; 3. increases the permeability of minor streets in order to reduce the longitudinal segregation (Figure 4).

As showed in Figure 4, the project reverses the hierarchy of transport modes, giving priority to sustainable forms of mobility.

A new configuration of roads is provided by narrowing the carriageway and extending sidewalks in favour of pedestrians, transit and cyclists' comfort and safety. To complete, public space retrieved from cars constitutes a single wide tape that in some points widens to integrate interstitial areas and wider surfaces, today not accessible. Some features, such as pavement surface, urban greenery, seating, and local services, orient people in moving through.

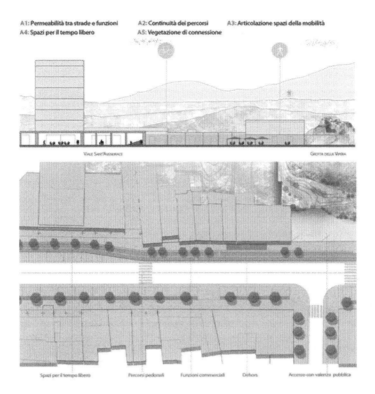

Figure 4. The new road configuration.

5 CONCLUSIONS

To summarize, the operational framework achieved three practical goals to activate processes of urban renewal: 1. consideration of the specificities of the context as important reference for the formulation of urban policies; 2. Acknowledgment and reinterpretation of the general types of actions, otherwise discontinuous and fragmented; 3. Activation of new practices in public administration processes for handling complex projects.

Specifically, the connective system outlines a new spatial and functional organization, that support a better quality of life in the neighbourhood.

REFERENCES

Belli A. (2006), "Cittadini e istituzioni: ascolto delle periferie e nuovi mestieri dell'urbanistica", in Belli A., (a cura di), "Pensare la Periferia", Cronopio, Napoli.

Congiu, T., Plaisant, A. (2018), The role of Connective Space in Regeneration, Urban Design 147, Streetscape, Urban Design Group Journal, London, ISSN 1750712X, pp. 18–20.

Salzano E. (2000), "Le Periferie cinquant'anni dopo". in F. Indovina (a cura di), "L'Italia è cambiata 1950-2000", a cura di F. Indovina, Franco Angeli, Milano.

Vergano A. (2015), "La costruzione della periferia: La città pubblica a Genova 1950-1980", Gangemi, Roma.

Pedestrians, Urban Spaces and Health – Tira, Pezzagno & Richiedei (eds)
© 2020 Taylor & Francis Group, London, ISBN 978-0-367-46171-3

In field assessment of existing pedestrian paths: A comprehensive approach towards pedestrian oriented neighbourhoods

S. Rossetti & M. Zazzi
Dipartimento di Ingegneria e Architettura (DIA), Università degli Studi di Parma, Italy

ABSTRACT: The paper aims at merging together previous studies and research activities developed by the authors on the geometry and on the accessibility levels of existing pedestrian paths, drafting the guidelines for a methodological approach to comprehensively assess the existing pedestrian network infrastructure at the neighbourhood level.

1   INTRODUCTION

Strictly focusing on the urban planning discipline fundamentals, walking should be the primary type of movement within the neighborhood urban unit, ensuring access to neighborhood services and facilities (see, i.a., Perry, 1929; Columbo, 1966), and also plays a crucial role in many contemporary urban planing movements and models, such as **Transit Oriented Development**, **New Urbanism**, **Smart Growth** and **Car-free cities**.

In addition to those approaches, already in 1986, the World Health Organisation (WHO) established the **Healthy Cities** programme, defining Healthy cities as cities that constantly create and improve the physical and social environments, investing resources to allow all users to perform health functions necessary for life.

Over time, the concept of *Healthy city* has been deeply intertwined to the one of **Active city** under the motto "A healthy city is an active city" (Edwards & Tsourous, 2008), and its more recent developments and declinations, like the *Active Design* approach, developed in the USA to contrast obesity by fostering physical activity and health in public spaces (see, i.a., Center for Active Design, 2010).

Among all those concepts and frameworks, the role of pedestrian mobility clearly emerges and dominates (see, i.a., ITF, 2012; Centre for Active Design, 2010; WHO, 2008): we can say that healthy and active cities are, first, **pedestrian oriented cities** and root on encouraging walking by providing adequate pedestrian infrastructures and promoting their use.

Therefore, the debate about creating pedestrian friendly environments and improving walkability is nowadays still dense and intertwined (see, i.a., Forsyth, 2015), and the research on how to proper design pedestrian paths is wide and articulated (see, i.a., VSS, 2009; ITF, 2012; Global Street Designing Initiative, 2016; Giuliani & Maternini, 2017).

But, which are the features that actually encourage walking within an existing neighbourhood? And how can we assess an existing pedestrian network and prioritise interventions?

2   OBJECTIVE OF THE PAPER

Within this framework, the paper aims at presenting a methodological approach for the in-field assessment of existing pedestrian networks at the neighborhood level, that can be considered as an holistic technique for evaluating pedestrian paths provisions and prioritise possible interventions. The proposed approach roots on in-field inspections of pedestrian paths, which are crucial lines of the road network: pedestrian paths represent specific and vulnerable sites on the roads, therefore they should be punctually assessed.

3 METHODOLOGICAL APPROACH

In the literature there are many walkability indexes and measures (generally based on urban form features like density, land use mix and street connectivity) used to describe the walkability level of a city, or of a given area (see, i.a., Forsyth, 2015; Conticelli et al, 2018). Those indexes mainly bases on ex-post evaluations made through GIS-based applications and geo-processing tools that process different datasets allowing to measure and assess the spatial walkability conditions, without detailed in field inspections.

The proposed methodology is slightly different, because it bases on in-field inspections that allow a comprehensive and very detailed and punctual approach to assess the existing pedestrian network infrastructure at the neighbourhood level. On the other hand, the proposed methodology is less speditive, and its replicability is affected by time needed for inspections and data entering.

Basing on the inspection table described in the previous paragraph, it could be possible to consider five areas of assessment, schematised in Figure 1, for each arc of the analysed pedestrian paths network. Those areas of assessment respond to specific questions that together constitute a comprehensive and holistic view of the features that contribute in creating pedestrian oriented neighbourhoods:

1) **Safety**: Which is the level of protection against accidents and risks for the pedestrians? Pedestrian paths and crossings should, first of all, ensure both road safety and perceived security of their users (see, i.a., Busi & Tira, 2001).
2) **Urban Accessibility**: Are urban functions and services of the neighbourhood connected and reachable within acceptable walking travel times? Pedestrian oriented neighborhoods not only have full pedestrian facilities such as sidewalks or paths, marked pedestrian crossings, appropriate lighting and street furniture, but should provide accessibility to the services, connecting neighborhood facilities and accomodating all the connectivity needs (see, i.a., Tiboni & Rossetti, 2014; Papa et al, 2018).
3) **Enjoyment**: Is the walking experience comfortable and attractive? In a pedestrian oriented environment, pedestrian paths may also include interesting and pleasant views and architectures, opportunities to sit, to stay and to meet other people, and services attractive to those who have other choices for getting around and getting exercise (see, i.a., Gehl et al, 2006).

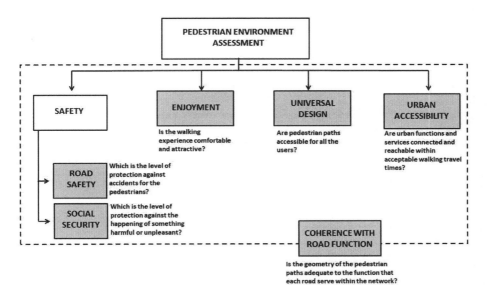

Figure 1.   The proposed approach for the assessment of the pedestrian network.

4) **Universal Design**: Are pedestrian paths accessible for all the users? Universal design includes not only accessibilty for the disabled, but focuses on all the vulnerable road users, which are not one entity, but a blend of different groups of people, with different character-istics, travel habits and behavioral patterns, having in common their difficulties to cope with motorised traffic (see, i.a., Pezzagno & Arenghi, 2017; Gargiulo et al, 2018).

5) **Coherence with the road function**: Is the geometry of the pedestrian paths adequate to the function that each road has within the network? To wrap up the proposed assessment, a last aspect should be evaluated to verify the different features previously investigated in relation to the fucntion of the road on which the considered pedetsrian path is located.

4   DESCRIPTION

To achieve those goals, a tool for punctual inspections of pedestrian paths was created, rooting on a bibliography review on the topic of pedestrian networks and developing previous experiences and models gathered by the authors in the field of slow mobility (see, i.a., Zazzi et al, 2018): it consists in inspection tables, to be filled in by technicians during on-site punctual inspections.

The proposed inspection sheet is an essential tool for highlighting both the quantitative and the qualitative features of each homogeneous arc of a pedestrian path. It is easy to apply and its standardization guarantees a smooth running of the inspections. It is quite comprehensive, so it allows further assessments, as proposed in the following paragraph. Figure 2 shows an extract of the adopted inspection table, filled in for an arc of via Paolo Toschi, in Parma.

5   RESULT OF THE RESEARCH

According to the proposed methodology, in field assessment of pedestrian paths are now under-going for the historic centre Parma, with a focus on its designation as Italian Capital of Culture for the year 2020. The specific work that is undergoing for Parma proposes to verify the accessi-bility of the urban environment of the historic centre of Parma through the analysis of the status of sidewalks, squares, public transport stops and entrances to places of culture. By defin-ing an accessibility map, the created model could be used in decision-making processes in urban policies, in the implementation phases of plans and in the definition of intervention priorities.

The approach could then be easily replicated to other context, and it is suggested its appli-cation at the scale of the neighborhood, to provide a comrehensive overview of the pedestrian environment.

6   CONCLUSIVE REMARKS

This type of analysis can provide the Public Administration a useful tool for assessing the accessibility of pedestrian mobility in our cities. At the end of the work it is possible to have a cognitive framework that highlights the critical aspects of the pedestrian paths analyzed, suggesting possible solutions to be implemented, and providing decision support both in the planning process (e.g. for Sustainable Urban Mobility Plans and sector plans for accessibility), and to the definition of intervention priorities for road maintenance works and public works.

However, it has to be clear that the process should find a balance between the number of data and informations collected and the real needs of the planning process. A detailed city-wide application of the whole approach is probably too time consuming in relation to the benefits obtained, therefore it may be suggested its application at the neighborhood level or for specific needs (e.g. for the case study of Parma, the approach is being applied to those pedestrian paths that connect places related to Parma 2020 events, therefore more likely sub-jected to pedestrians and turists flows in the near future). Anyway, the inspection sheet can be taylored to specific needs and being reduced or adapted according to specific objectives.

Historic center	Identification code	P_1	Map
	Viale Paolo Toschi		
Year: 2019	Quality index		

GENERAL	
TYPE	☒ Sidewalk ☐ Virtual sidewalk ☐ Bike and pedestrian shared track ☐ Path within green areas ☐ Pedestrian crossing ☐ Shared space
PREDOMINANT LAND USE This field shows the prevailing feature of the buildings overlooking the pedestrian path. If there are multiple functions, overriding function means the function that takes up most of the volume of the buildings overlooking the route.	☐ Prevailing residential use ☒ Prevailing commercial/tertiary use ☐ Prevailing productive use ☐ Prevailing services/facilities use ☐ Agricultural area
SPEED LIMIT	☐ Speed limit 30 km/h ☒ Speed limit 50 km/h ☐ Speed limit 70 km/h ☐ Street intended solely for pedestrians or cycles
PROTECTION This field records the level of pedestrian protection, passing in the path, relative to the roadway. The identification of the level of protection on the part of the detector is possible by selecting three levels thus determined: -No Protection: there is no distinction between the footpath and the roadway; -Low protection: presence of the only sidewalk -Average protection: there are additional physical elements of discontinuous protection (such as trees, bollards, barriers, pedestrian barriers, etc.); -High protection: physical security elements are continuous.	☐ No ☐ Low ☒ Average ☐ High

Figure 2. An extract of the proposed inspection table (page 1 of 6) (Inspection and data entry by Maddalena Moretti).

Being able to have a real image with precise information on the accessibility of pedestrian paths would allow researchers and local administrations to identify which areas need integration, modifications or additions to guarantee a safe and pleasant walking experience to all the users, thus implementing localized interventions, and avoid the inclusion of redundant and invasive solutions or tools.

REFERENCES

Busi R., Tira M. (2001). Safety for pedestrians and two-wheelers. Bios, Cosenza.
Center for Active Design (2010), Active Design Guidelines, New York.
Columbo V. (1966). La ricerca urbanistica. Giuffrè, Milano.

Conticelli E., Maimaris A., Papageorgiou G., Tondelli S. (2018). Planning and Designing Walkable Cities: A Smart Approach. In: Papa R., Fistola R., Gargiulo C. (eds) Smart Planning: Sustainability and Mobility in the Age of Change. Green Energy and Technology. Springer, Cham.

Edwards P., Tsourous A.D. (2008). A healhty city is an active city: a physical activity planning guide. WHO Europe, Copenhagen.

Forsyth A., (2015). What is a walkable place? The walkability debate in urban design. Urban Design International 20, no.4: 274–292.

Gargiulo, C., Zucaro, F., & Gaglione, F. (2018). A Set of Variables for the Elderly Accessibility in Urban Areas. TeMA - Journal of Land Use, Mobility and Environment, 53–66.

Gehl, J., Gemzøe, L., Kirknæs, S., Søndergaard, B. S. (2006). New City Life, The Danish Architectural Press, Denmark.

Giuliani F., Maternini G. (eds) (2017). Percorsi Pedonali. Progettazione e tecniche di itinerari ed attraversamenti. Egaf, Forlì.

Global Designing Cities Initiative (2016). Global Street Design Guide. Island Press.

ITF (2012), Pedestrian Safety, Urban Space and Health, ITF Research Reports, OECD Publishing, Paris.

Papa, E., Carpentieri, G., & Guida, C. (2018). Measuring walking accessibility to public transport for the elderly: the case of Naples. TeMA - Journal of Land Use, Mobility and Environment, 105–116.

Perry C. (1998). The Neighbourhood Unit (1929) Reprinted Routledge/Thoemmes, London, p. 25–44.

Tiboni, M., Rossetti, S. (2014). Achieving People Friendly Accessibility. Key Concepts and a Case Study Overview. TeMA - Journal of Land Use, Mobility and Environment, Special Issue, 941–951.

VSS (2009). SN640070. Trafic piétonnier; norme de base, Association Suisse des professionnels de la route et des transports, Zurich.

Pezzagno M., Arenghi A. (2017), Disabilità negli itinerari pedonali. In Giuliani F., Maternini G., Percorsi Pedonali, Egaf, Forlì, 147–194.

Zazzi M., Ventura P., Caselli B., Carra M. (2018). GIS-based monitoring and evaluation system as an urban planning tool to enhance the quality of pedestrian mobility in Parma. In Tira M., Pezzagno M. (eds.), Town and Infrastructure Planning for Safety and Urban Quality: Proceedings of the XXIII International Conference on Living and Walking in Cities, CRC Press, Taylor and Francis group, London, pp. 87–94.

Pedestrians, Urban Spaces and Health – Tira, Pezzagno & Richiedei (eds)
© 2020 Taylor & Francis Group, London, ISBN 978-0-367-46171-3

A citizen science approach to assess the perceived walkable environment and identify elements that influence pedestrian experience at the University of Malta

C. Cañas
Institute for Climate Change & Sustainable Development, University of Malta, Malta

M. Attard
Department of Geography, University of Malta, Malta

M. Haklay
Department of Geography, University College London. UK

ABSTRACT: The aim of this ongoing research is to develop and test a participatory method based on citizen science principles to systematically assess perceived walkability and identify the elements of the walkable environment (WE) that influence pedestrian experience. To do so, an empirical study at the University of Malta Campus will be conducted, where volunteers can significantly contribute to the research by collecting georeferenced subjective and objective observations of the WE as they go about their daily routines. Such data will be spatially aggregated and statistically analised to assess the degree and the spatio-temporal distribution of the perceived walkability within the study area. Once the study is concluded, a data quality assessment and the interpretation of the research outcomes will determine the suitability and effectiveness of this innovative approach.

1 INTRODUCTION

Many efforts to promote walking have focused on studying the walkable environment (WE) to identify determinants that enable and encourage it (Saelens & Handy, 2008). Thus, a well-established conceptual definition of the WE –also referred as the walkability of a place– would provide a solid reference to assess and study its relations with other variables, and therefore guide research and policies (Forsyth, 2015; Lo, 2008). However, as walking is a ubiquitous activity with implications in many fields, different concepts of walkability have been developed to suit different studies (Fitzsimons, 2013).

One of the most prevalent approaches to defining walkability relates to the premise that urban density, land use diversity and network design, which determine pedestrian proximity to destinations, influence walking (Ewing & Cervero, 2010). Consequently, walkability is commonly used as a composite variable based on objective observations of urban characteristics, which arguably became the most used approach to define, operationalise and assess walkability since the 2000s (Frank et al., 2005).

However, while this approach seemed adequate in studies for utilitarian walking, it often failed to fully explain and predict walking as recreational or socio-economic activity (Kang et al., 2017), linked to some research on liveability and urban sustainability, which suggest that the WE cannot be abstracted from its social setting and should be assessed through the citizens who experience it (Bornioli, 2018). Since then, some walkability research has identified more subjective constructs to define the WE, such as safety, comfort, pleasantness or vibrancy, amongst others (Alfonzo, 2005).

This challenges previous walkability studies solely based on objective observations, where the WE is defined through tangible elements that can be empirically measured and tested through scientific-method approaches (Park, 2008). By contrast, pedestrian-centred walkability research based on subjective experiences require new ways of observing, measuring, analysing and interpreting the WE. Recognising this challenge, an emerging research approach known as citizen science (Cavalier & Kennedy, 2016) can be applied to set a participatory methodological framework where pedestrians significantly contribute to the research.

2 RESEARCH APPROACH

The first part of the research involves the development of a framework combining theoretical principles of walkability research -to establish subjective constructs that define and operationalise the perceived WE-, and methodological aspects of citizen science -to ensure a cost-effective and in-depth participatory walkability assessment-, for the observation, measurement and assessment of the WE.

The theoretical framework evolves from socio-ecological walkability studies, which recognise that walking is not only the result of rational reasoning and interaction with the environment, but also to personal mental states and individual perceptions (Le Vine et al., 2013). The need for categorising all the multiple factors affecting walking behaviour and their relative impact, led Alfonzo (2005) to develop a model of 'walking needs', proposing a hierarchy of environmental criteria for pedestrians. She suggested that in order for a person to choose to walk, the main determinants are feasibility, accessibility, safety, comfort and pleasantness. This framework explores how environmental characteristics are perceived to contribute towards positive experiences that affect people's decision to walk and identifies environmental factors that fulfil such needs. Similarly, other studies added new pedestrian needs as walking behaviour factors, such as usefulness, enjoyment, and sense of belonging (Mehta, 2008; Mateo-Babiano, 2012). Accordingly, the theoretical construct that defines perceived walkability in this study is based on the concepts of safety, comfort, peasantness and vibrancy.

The methodological framework relies on citizen science principles (ECSA, 2015), which lay the basis to develop, implement and evaluate comprehensive participatory research. Although participatory practices have been long used in walkability studies, citizen science differs to simply volunteer to participate in focus groups or respond to pedestrian surveys and interviews. Particularly, this study embraces certain citizen science concepts that define its research design.

Firstly, the transformative potential of citizen science applied to walkability engages and empowers pedestrians with the necessary tools and platforms to raise awareness about their needs, which may influence policymaking and improve their lifestyle (Ancianes, 2011). And secondly, the implementation of online environments allow constant interaction among participants and researchers, guarantees data collection at the necessary spatio-temporal scale, and provides in-depth insights that otherwise would be unfeasible to collect or go unnoticed. The ubiquitous use of smartphones with constant internet connection and built-in sensors, along with customised mobile apps and web-based platforms have revolutionised collaborative research. In this line, this research explores the use of social media platforms as services for data collection, management and visualization, as well as a channel of communication. As a result, pedestrians can voluntarily contribute to the research as they go about their daily routine.

3 DESCRIPTION

This pedestrian-centred walkability assessment combines georeferenced subjective and objective observations of the WE, collected by pedestrians via social media using their smartphones, as shown in *Figure 1*. On one hand, subjective data on perceived walkability is collected using posts containing qualitative dichotomous variables, which express either positive or negative

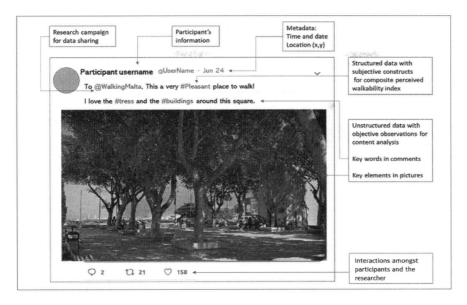

Figure 1. Data included in participant's post. Drawn by author.

experiences based on the concepts of safety, comfort, pleasantness and vibrancy. This method provides a structured data collection method for a self-reported survey where participants rate the WE through the variables "safe/unsafe, comfortable/uncomfortable, pleasant/unpleasant, and vibrant/dull". On the other hand, objective data on the WE is collected through attached pictures or comments. This allows participants to identify the presence or absence of elements of the WE considered relevant for their walking experience.

All data are then processed and analysed, resulting in two main research outcomes. Firstly, a perceived walkability index (PWI) is developed to assess the degree and spatial distribution of pedestrian perceptions. Social media posts can be spatially aggregated and analysed to calculate a composite score from the variables describing pedestrian experience. A weighted average formula calculates the percentage of positive perceptions out of total observations for each variable (safety, comfort, pleasantness and vibrancy), to then weight them based on their frequency, and finally aggregate them all into a single value determining the perceived walkability for that certain place and time. Secondly, the linked objective observations on the WE can be statistically analysed in relation with the perceived walkability to identify the most relevant elements of the WE that may positively or negatively influence their experience.

4   EXPECTED RESULTS AND CONCLUSIONS

While results from the planned empirical study are expected to show the spatial distribution of perceived walkability and the elements of the WE that influence pedestrian experience within the study area, the quality of the resulting data will determine the suitability and effectiveness of this approach once the empirical study takes place and the research is concluded.

Nevertheless, this exploratory research can be considered a well-suited approach for investigating hidden insights behind the complex, interrelated, and multifaceted social processes in the assessment of the perceived WE. Especially to some emerging concepts, such as walkability pleasantness and vibrancy. It may also contribute to theory construction, and help uncover interesting and relevant questions and issues for walkability follow-up research and policies.

REFERENCES

Anciaes, P.R., (2011). Urban transport, pedestrian mobility and social justice. A GIS analysis of the case of Lisbon Metropolitan Area. Doctoral Dissertation. University of London.

Bornioli, A., (2018). The influence of city centre environments on the affective walking experience. Doctoral dissertation. Centre for Transport and Society, Faculty of Environment and Technology, University of the West of England, Bristol.

Cavalier, D. and Kennedy, E.B., (2016). The Rightful Place of Science: Citizen Science. Tempe: Consortium for Science, Policy & Outcomes.

ECSA., (2015). Ten principles of citizen science. The European Citizen ScienceoAssociation.lRetrievedlfroml https://ecsa.citizenscience.net/site/default/files/ecsatenprinciplesofcitizenscience.pdfl(accessedl25. Jul 2019).

Ewing, R. and Cervero, R., (2010). Travel and the built environment. Journal of the American Planning Association, 76(3) pp. 265–94.

Fitzsimons D. L., (2013). A multidisciplinary examination of walkability: Its concept, assessment and applicability. Doctoral dissertation. Dublin City University. School of Health and Human Performance.

Forsyth, A., (2015). What is a walkable place? The walkability debate in urban design. Urban Design International, 20(4), 274–292.

Frank, L.D., Schmid, T.L., Sallis, J.F., Chapman, J.E. and Saelens, B.E., (2005). Linking objectively measured physical activity with objectively measured urban form: findings from SMARTRAQ. American Journal of Preventive Medicine, 28(2), pp. 117–125.

Kang, B., Moudon, A.V., Hurvitz, P.M., Saelens, B.E., (2017). Differences in behavior, time, location, and built environment between objectively measured utilitarian and recreational walking. Transportation Research Part D: Transport and Environment, 57, pp. 185–194.

Le Vine, S., Lee-Gosselin, M., Sivakumar, A. and J. Polak. (2013). A new concept of accessibility to personal activities: Development of theory and application to an empirical study of mobility resource holdings. Journal of Transport Geography, 31, pp. 1–10.

Lo, R., (2009). Walkability: what is it? Journal of Urbanism: International Research on Placemaking & Urban Sustainability, 2(2), pp. 145–166.

Mateo-Babiano, I., (2012). Public life in Bangkok's urban spaces. Habitat International. 36 (4), pp. 452–461.

Mehta, V., (2008). Walkable streets: pedestrian behaviour, perceptions and attitudes. Journal of Urbanism, 1, pp. 215–245.

Park, S., (2008). Defining, Measuring, and Evaluating Path Walkability, and Testing Its Impacts on Transit Users' Mode Choice and Walking Distance to the Station. Doctoral Thesis. University of California Transportation Centre.

Saelens, B.E. and Handy, S., (2008). Built environment correlates of walking: A review. Medicine and Science in Sports & Exercise, 40 (7), pp. 550–566.

Pedestrians, Urban Spaces and Health – Tira, Pezzagno & Richiedei (eds)
© 2020 Taylor & Francis Group, London, ISBN 978-0-367-46171-3

Freewheeling thoughts about public space

M.R. Ronzoni
Università degli Studi di Bergamo, Italy

ABSTRACT: Starting from the definition of public space, this paper focuses on the changing meanings and functions of public space over time and it tackles the following questions: what is public space today, where and how can we enhance it, regenerate it, and integrate it? The context investigated is the municipal area of Bergamo, where a section of the transversal municipal territory that goes from the historic boroughs of the lower city to the extreme municipal periphery was examined. If the objective is to enhance these spaces and regenerate them, it is necessary not only to locate them, but also to relate them among one another and to ensure their accessibility, continuity and above all visibility. This is made possible by employing the maps that we have called Usage Maps. These maps help identify the degree of use of public spaces and their potential. Also, they help the population perceive public spaces for the role they currently have and can take up in the future, use them more effectively and with greater satisfaction, live them and live them for a longer time.

1 INTRODUCTION

Today we think of public space in very different way from how the concept of public space was perceived in the past.

At times, however, we can observe a circularity in the way we use public space: for instance, the space devoted to the market in the middle ages (market place) became the space for the car in the second half of the last century and it is now again the space for the market.

Physical public space may in fact be quite different from public space as perceived by users. In the past public space belonged to the community. Today we have public spaces that are in use by people, but that are private property. These spaces attract crowds of people. I am thinking, for instance, of the Markthalle in Rotterdam, the Potsdamer Platz in Berlin or Elbphilarmonie in Hamburg. However, while traditional public space as such allows the user to appropriate it (as public space belongs to everyone, and therefore it is also mine), in the above cited cases the sense of identity of the place is lacking: you cannot appropriate it because it belongs to someone else, you can only use it without owning it.

On the other hand, this is in line with our times. Recently, for instance, the Swedish furniture manufacturer IKEA announced plans to start renting some of its pieces of furniture: you will use them and, when they become obsolete or deteriorate, IKEA will take them back, refurbish and resell them, thereby extending the life of the products. The same thing happens with public space.

Piazza San Marco in Venice attracts crowds. Here the past canons of public space coexist in a contradictory fashion with the way of enjoying today's public space. The need to amaze and to appear rather than to be, often resulting in the organization of events of huge appeal, obliges the local administration to limit access to the square to 20,000 people by privatizing, in a certain sense, the public space. In some cases, where public space fails to convey such a sense of identity, the public place can be hardly experienced and, consequently, it becomes dangerous.

I have studied this topic in some theses[1] that I have promoted in the Bergamo area, exploring both parts of the built environment as well as peripheral areas.

2 DEFINITION OF PUBLIC SPACE

These theses works have shed light on some elements that it is useful to investigate when discussing about public space. On the one hand, there is the definition of public space as a shared place, a meeting place, a dialogue place. Let us think of the square as the public place of excellence but also a place where certain needs (that in the past we would have struggled to satisfy) are met. An example can be represented by the neighbourhoods of the Modern Movement, where the concept of "Existenzminimum" housing is found. The minimum accommodation is an essential private space, it represents a way of experiencing accommodation by optimizing the usability of the spaces. The essentiality that characterizes the minimal housing has led the designers of the time to introduce a solution still widely employed in Northern Europe: that of the common accommodation, i.e. the shared space, made available to the inhabitants of the building/neighbourhood, that enables the activities that do not fit within the limited size of the apartment. I refer to a space to read, to meet friends, for the unbridled play of children. This is also the case of public space. The park is the back yard we would all like to have, where to stay in contact with nature, where to relax, where to take care of our body and soul. The square is the place where to get together and celebrate happy moments, but also to express dissent or simply to meet and start a dialogue. Not everything is allowed or possible to us as private subjects. Public space should take care of ensuring that at least part of what we cannot afford becomes accessible to us, of course, by means of a significant common effort.

In order to define public space, I can first classify it as public property or private property with public use. Both can be open spaces or enclosed spaces. In the past – let us go back to nineteenth and early twentieth century maps – only squares and streets were marked on maps: those were the only public sphere. The private sphere was dominant. Those who could afford it had ample living spaces where they could exercise their pleasure and follow their own inclinations. The others, the regular people, had nothing. Little by little, society has become aware of these needs and has given different faces to their satisfaction through new definitions of public space. Then followed the economic aspects. At the outset, the public was actually public. With time, due either to excessive costs (that the community is less and less able to cover) or to potential profits (that public space may also exploit as a marketable place of aggregation), solutions for public use but private initiative started to be envisaged. Chorus Life, to remain in the Bergamo area, constitutes for example an interesting reference that introduces a new formula of choral life entirely controlled by the private sector.

This investigation starts from the definition of public space. It then focuses on the migration of meaning and functions of public space over time and it finally tackles the following issues: what is public space today, where and how can we enhance it, regenerate it, integrate it?

3 USAGE MAPS AS TOOL OF REVITALISATION

In order to evaluate public spaces, I attach a number of survey charts aimed at framing the territory under investigation and the dynamics that characterize it. It would be very helpful to consider the complete sequence of charts prepared for the theses' research. It would however take too long and so I prefer to recall only the main results.

---

1. thesis graduates "Gli spazi pubblici nei quartieri di Borgo Sant'Alessandro, Borgo San Leonardo, Santa Lucia, San Paolo, Loreto e Longuelo di Bergamo, come parte strutturante del tessuto urbano", written by Simone Vismara, Academic Year 2013/14, supervisor Maria Rosa Ronzoni and thesis graduates "Il ruolo del pubblico nella caratterizzazione degli spazi periferici", written by Nicola Zana, Academic Year 2014/15, supervisor Maria Rosa Ronzoni

Mapping both public spaces and privately owned spaces used as public spaces is essential. The analysed context rests within the municipal area of Bergamo. The first step concerns the definition of the field on which to conduct the investigation, the function assumed by the analysed space, the consistency, the time of use of existing public spaces. It was decided to examine a section of the transversal municipal territory that goes from the historic boroughs of the lower city to the extreme municipal periphery and to collect a series of data and information useful to establish its functioning. (Figure 1, Figure 2) This also serves the purpose of giving shape to a system of public spaces in relation to one another, able to favour continuity and to overcome the fragmentation effect that is often associated with the image of urban additions that have taken place especially in the second half of the twentieth century. On and amongst those public spaces, work has been done to develop design solutions that favour soft, especially pedestrian, mobility. If the objective is to enhance these spaces and to regenerate them, it is necessary not only to locate them, but also to relate them to one another and to ensure their accessibility, continuity and above all visibility This is made possible by employing the maps that we have called Usage Maps. (Figure 3) These maps help identify the degree of use of public spaces and their potential. Also, they help the population perceive public spaces for the role they currently have and can take up in the future, use them more effectively and with greater satisfaction, live them and live them for a longer time. It is necessary to fit into such a scheme with a project that integrates and enhances what is already present. In the territory of the municipality of Bergamo, the work of enhancing public spaces and their systemization finds an important reference in the green public areas, represented among other things by the areas included in the Parco dei Colli di Bergamo and by the Green Belt Project elaborated in the PGT[2]. Among the data collected we find information on the population, the distribution of the population by age group. It is also important to collect data on the functions available in the neighbourhood and on their percentage distribution. The data on the population density has the purpose of documenting the most densely populated areas and developing some

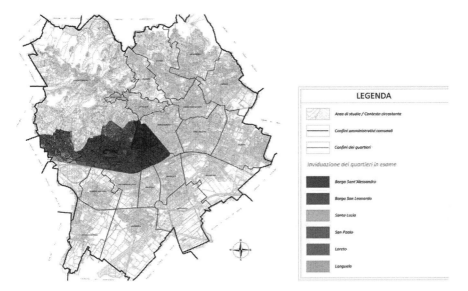

Figure 1. Neighbourhoods analyzed in the thesis work by Simone Vismara, distributed from the center to the periphery. They have been chosen to provide a transversal and diversified reading of the equipment and consistency of public space.

---

2. PGT stands for Piano di Governo del Territorio and it is an urban plan introduced in Lombardy.

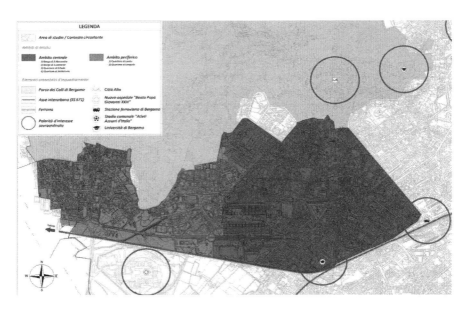

Figure 2. Neighborhoods analyzed in the thesis work by Simone Vismara, represented for the area in which they fall: central (in red) or peripheral (in blue). Polarities of higher rank are also indicated.

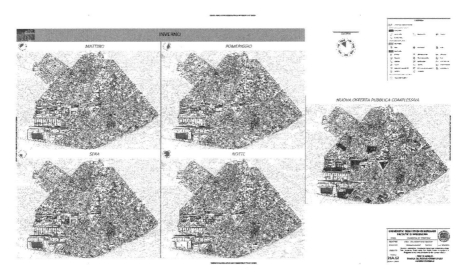

Figure 3. Map of the degree of use elaborated in the thesis of Simone Vismara. The study was conducted for the four seasons for the different days of the week and for the different time slots. The example given refers to a Sunday in the winter season, and shows the situation of the use of public spaces in the different time slots.

indications in relation to the public spaces detected in these areas and their degree of use. As expected, the population density is decreasing in the transition from the central areas to the more peripheral areas. Still with regard to demographic factors, it is useful to represent the population by age group. This representation allows to match the regeneration interventions of public spaces with the trends corresponding to the prevailing ages. An additional information appears useful in redefining public space: the representation of functions. Here we find that the historic districts are characterized by a balanced functional mix, whereas the more peripheral and more recent areas exhibit a prevailing mono-functional use. Below, by way of example, we show some diagrams extracted from the research.

4 DEGREE OF USE OF PUBLIC SPACES

The tables were designed to represent the degree of use of public spaces. The degree of use of a public space can be defined as the integrated combination of the times within which the public utility service is offered, and the level of employment by the users to whom the service is addressed. Intuitively, the study process of the degree of use thus defined consists in indicating clearly and unambiguously how each space functions during the hours that make up the days of the typical week, in turn distinguished in relation to the reference seasonal period. The elaboration of these maps guided the subsequent design choices that went to "turn on" the studied territory, enhancing and integrating the existing offer.

5 AN IMPORTANT INDICATOR

Among the results of the research I wish to drive attention to an indicator, the one related to the ratio between the surface of pedestrian areas and the total surface of roads and streets within the territory under investigation. As mentioned, the work on the redevelopment and enhancement of public space necessarily passes through a redevelopment of the path system. Before the intervention this ratio is equal to 0.9% while, following the redevelopment of pedestrian connections, a potential 4.2% is reached. All this translates not only into a greater extension of paths able to favour the relations between the different boroughs and the system of the public offer, but, above all, into a concrete push towards the re-appropriation of open public space as a place of life for the city and its inhabitants.

This means a prolonged and extended use of spaces already available, but not perceived as such, or a new use of spaces now present but not used for the specific purposes.

6 CONCLUSIONS

Listening to places, both in the expression of their inner being and through the voice of their inhabitants, is certainly helpful in defining living spaces that better meet the needs of the inhabitants.

Certainly the current situation is such that it becomes increasingly easy to access sensitive data on the way cities are used, on the prevailing flows, on the most popular aggregation points. Processing this additional information, and redirecting the design of the way of using and optimizing the spaces of the neighborhood, is certainly an objective to aim for in the future development of research. In the simplicity of the method presented, this research is easily replicable in every context, presumably equal in the the data to be collected and different in the outcome of the observation and the solutions to be proposed.

REFERENCES

Feireiss Kristin, Hamm G. Oliver. (2016). Transforming cities. Jovis Verlag. Pages 1–192.
Ferguson Francesca, (2014). Make shift city. Jovis Verlag. Pages 1–254.
Haidn Barbara, (2016). Dempolis, das Recht auf oeffentlichen Raum. Park Books. Pages 1–287.
Lonafi Kamel, (2016). Public Spaces. Jovis Verlag. Pages 1–176.

Green infrastructures / Public transport

Green infrastructures for urban and territorial regeneration. The sustainability of the contemporary city of Somerville

S. Cioci
University of Study "Sapienza", Rome, Italy

ABSTRACT: The elements that put our planet in crisis derive to a large extent from the increasing phenomena of human anthropization. Identifying, promoting and preserving a strategically planned Green Infrastructure network (GI) can provide ecological, economic and social benefits. Through the method replicable and flexible, the quantification of the environmental and economic benefits applied on the plans adopted by the Inner Core governances of Boston, the research aims to demonstrate how strategic green infrastructure can lead to territorial adaptation to climate crisis and for smart and sustainable growth of the entire region.

1 INTRODUCTION

The research path reported in the paper aims to take the best we can of the scientific studies currently available with the aim, not so much to gather information or satiate our curiosity but to become aware of the current global climate crisis and identify which it is the contribution that everyone can make.

Through reckless exploitation of nature, the environmental and ecological crisis is a dramatic consequence of the uncontrolled activity of the human being that risks, moreover, to be itself a victim of such degradation. The urgency and the need for a radical change in the conduct of our lifestyles affect not only the rethinking of our cities but also, and above all, an authentic social and moral progress. For this reason, many governances are preparing for an "ecological conversion" capable of eliminating the structural causes of the dysfunctions of the world economy and correcting the growth models that seem unable to guarantee respect for the environment. The increase in vegetated surfaces is increasingly recognized as a resource both in terms of climate change mitigation and as a tool to promote urban quality, indeed, in 2015 the World Economic Forum (WEF)[1] included the increase in the Global Agenda Council of greenery as the first point among urban initiatives, stating that: "Cities will always need large infrastructure projects, but sometimes small scale infrastructure from cycle lanes and bike-sharing to the planting of trees for climate change adaptation, can also have a big impact on an urban area".

Associated with this climate crisis, the loss of control by the government of the territory, highlighted the inability to curb issues such as territorial fragmentation and the consequent urban decay. "The result is a deterioration of the landscape, perforated and consumed in its parts, which has multiplied the types and number of voids found in the contemporary city-territory" (Salzano, 2012).

The new visions that characterize the international scenarios, the new paradigms that guide the conservation of nature, the reticular perspectives that are looming in contemporary cities and

---

1. The World Economic Forum is the international organization for public-private cooperation. Founded in 1971 as a non-profit association based in Geneva, Switzerland, the Forum engages the largest political, entrepreneurial and company leaders in shaping global, regional and sector agendas.

territories, call for complex and multifunctional, trans-scalar and multi-sectoral sustainable strategies. The instrument of this new approach to the territory is certainly the Green Infrastructure (GI), a terminology coined in the United States in the mid-1990s (Henein & Merriam, 1990) which highlights the importance of the natural environment in territorial and urban planning.

2   THE PURPOSE OF SMART AND SUSTAINABLE GROWTH

The design of the continuous fabric of open spaces aimed at counteracting the territorial fragmentation, the reuse of abandoned places, and restitching of the neglected areas, constitute the main challenges for the project of the contemporary city that with a sustainable and environmentally friendly key, sets the final goal of regaining the green open spaces, observing them as places of the identity for the communities. Configured initially as an ecological network system for the enhancement of its ecosystem services, the GI described here, aims to demonstrate how, in the Boston metropolitan area in Massachusetts, this multifunctional planning strategy has expanded its functions as an integration tool for the smart and sustainable growth of Inner Core urban areas. From the provision of ecosystem services such as the supply of food and water, to the regulation and improvement of air quality, from the support of nutrient cycles to the supply of cultural services and to environmental justice[2], the GI is affected by numerous concepts able to realize the notion of *Triple Bottom Lines*: People, Planet, and Profit (TBL or 3BL)[3] which, precisely, summarizes the wide range of advantages from the three sustainability "Es": Environmental, Economy, and Equity, that offer an effective and functional framework to define the benefits of the green infrastructure.

3   METHODOLOGICAL APPROACH AND RESEARCH APPROACH

The aspects mentioned are only the preliminary perspective for the motivation of field research and the choice of the case study in the Boston metropolitan region. Through the analysis of data on future forecasts of the climate crisis, the GIS (Geographic Information System) monitoring of historiographical processes, the environmental and socio-cultural context, and the strategic visions of the plans adopted by the governance, it was possible to identify the places that, o due to historical developments or environmental change, they must be rethought in a tactical perspective capable of uniting the needs of citizens with the need to adapt to the climate crisis. With the aim of demonstrating the effective effectiveness in the adoption of Best Management Practices (BMPs)[4] strategies, the research, therefore, focused on the analysis and evaluation of the economic benefits deriving from the provision of ecosystem services provided by the urban green infrastructure for the city of Somerville. The replicable and flexible method is defined by a coherent sequence of two phases which consist of

2. The principle that inspired movements and organized groups for the defense of civil rights that recognize the environment as an element of equity and social justice, the concept of environmental justice has been affirmed in the United States since 1980s. The sociologist Robert Bullard, known as the father of environmental justice, is the first to demonstrate how the distribution of risks and environmental damage is based on the identification of particularly weak territorial contexts from the social, economic, political point of view such as those inhabited by poor communities, from African Americans, Hispanics, Native Americans who, precisely because of their weakness, are less likely to oppose or more inclination to yield to environmental "blackmail": environmental degradation in exchange for jobs.
3. Coined by John Elkington in 1994, the triple bottom line is made up of notions of social, economic and environmental equity (Elkington, 1998).
4. From the management of land sealed from urbanization to the redevelopment of areas that suffer from the lack of environmental justice, BMP sustainable rainwater management systems and mitigation of high surface temperatures include a wide range of renaturalization practices such as: the use of green roofs, the porous floors, the increased vegetation and the bioretantion and infiltration systems such as rain gardens and boumpouts.

measuring the environmental benefits provided by every single green practice and the subsequent assignment of the economic value, where possible, of these benefits.

4 DESCRIPTION

4.1 *Innovative aspects*

The profound rethinking of urban planning and protection of the territory in anticipation of the devastating effects of the climate crisis has led the cities of the Inner Core of Boston to adopt strategies of prevention, defense, and adaptation of both natural heritage and urban areas. Effective instrument to obtain the ecological, economic and social benefits required for smart and sustainable growth, the LandLine plan of Greater Boston, was studied by the Metropolitan Area Planning Council (MAPC), and approved in 2018, with the aim of realizing green infrastructure able to reconnect, through a system of hubs and links, the landscapes of high environmental quality that are present throughout the territory with the densely urbanized spaces of the Inner Core. By guaranteeing continuity in the landscape that exploits the existing ecological networks and the new greenways and footrail routes programmed, the MAPC has designed a complex of networks between peri-urban natural heritage and urban outlying areas, configuring the new green "framework" for landscape's resiliency and the territorial's sustainability (Figure 1.).

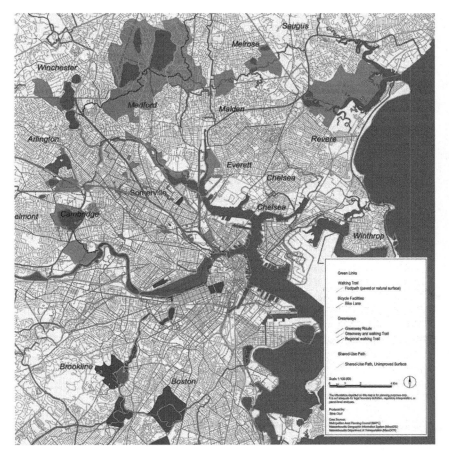

Figure 1.   MAPC LandLine plan for the regional reconnection of the Inner Core of Boston, Massachusetts. Source: Silvia Cioci for MAPC.

This ecological-functional approach, necessary to guarantee the continuity of the ecosystem processes and the minimization of the phenomena of fragmentation and soil consumption, places communities, and the benefits that these can derive in a good environment, at the center of research. In this scenario, the cities of Boston, Cambridge, and Somerville, key players of the international innovative scene, have drawn up urban management plans and programs that see in the BMPs renaturalization practices, the tools capable of achieving the 3BL objectives and the smart and sustainable growth of the whole region.

4.2 *Research and/or technical developments*

It is no secret that Somerville, defined until a few years ago "the little city that could", is today the city that is making significant changes both in terms of re-planning of the urban fabric and on the scene of regional modernization. Through a decision-making process that involved the entire community, the city adopted in 2009 the SomerVision plan that includes more than 584 objectives, values, policies and actions ranging from urban environmental sustainability to the creation of new open spaces, from the development of areas transformable to the increase in public transport, from the encourage of new jobs creation to the construction of new low-cost housing. Among the proposals for the urban revival of the city of Somerville, the green infrastructure was the decisive strategy to satisfy different functions.

Demanding a procedural approach where voids are offered as cells for "colonization" (La Greca & Carta, 2017), of a willing society to a new relationship with the natural world (Sykes & Brown, 2015), the policy of de-densification foreseen in the Zoning Overhaul of Somerville, has arranged interventions for the modification of the use destinations on public and private spaces, allowing, thus, to rethink those voids that for years had been forgotten.

5   RESULT OF THE RESEARCH

Estimates of ecosystem services are included in a collection procedure and the analysis of standardized data using environmental archives applies to the entire Inner Belt area of Somerville (US EPA, 2014). Predicting first a redevelopment on a 1 hectare local area of intervention in the town of Somerville, the study provided a detailed description of the estimates of four urban ecosystem services: (i) Regulation of water controls; (ii) energy savings; (iii) local removal of atmospheric pollutants; (iv) carbon sequestration. In the second phase, the identification of the economic values for each benefit provided by the different strategies was established on the basis of the environmental resources identified in the previous step and then translate into monetary amounts depending on the relevant benefit category. According to the selected scenario, the use of BMPs shows a reduction of the impermeable surface of about 21.5%. From this data, it was possible to determine the replacement of water pollutants such as suspended solids (TSS) and phosphorus and the consequent moderation of the use of water purifiers. The employment of green infrastructure strategies for urban renaturalization has also shown how increased vegetation can reduce the use of air conditioners and heat pumps by ensuring, thus, energy-saving both in winter and summer. With the aim, moreover, of verifying the effectiveness of the multifunctional strategy in mitigating climate change, the study has provided optimal results on the local removal of atmospheric pollutants and CO_2 storage, highlighting an emission removal gain of approximately 8 tonnes per year. The environmental multifunctionality demonstrated in the research carried out, although already considerable, has not been limited only to the aspects of mitigation, adaptation, protection, and environmental justice, the study has shown, indeed, as the use of green infrastructure strategies guarantee a not small economic gain allowing, for 1 hectare of land, a monetary return of up to USD 25,000.00 per year (Cioci, Eckelman, & Onnis-Hayden, 2019).

6 CONCLUSIONS

The strategy envisaged by the MAPC and the governances of the Inner Core in Boston invest, at least, three dimensions of urban resilience: the environmental dimension, guaranteeing security in terms of adaptation to climate change through systems that help mitigate flooding phenomena, to storage of air pollutants and to reduce of the urban heat island effects; the economic dimension, preferring planning that takes advantage of nature's ability to screen, store and alleviate environmental events that increasingly force the use of energy resources and simultaneously produce an attractive urban environment for new businesses; the social dimension, central to the ecological conversion of urban areas, capable of stimulating a paradigm shift in urban metabolism based on the recycling of resources and on a social and identity re-appropriation of common goods.

A subject of considerable interest in the project of the contemporary and innovative city, the green infrastructure, interpreted as a vegetal system of reconnection between urban spaces and the natural resources of the Boston metropolitan region, represents an example of how it is possible to enhance typical spaces of peri-urban and extra-urban areas but also of urban voids, placing, at the same time, a limit to the fragmentation of environmental habitats and contributing to the protection and safeguarding of biodiversity and ecosystems. Moreover, where the psycho-physical stresses linked to daily lifestyles can only be amplified in the face of future forecasts of the imminent climate change, the study leads to the attention on the capacities that the green infrastructure can provide in reducing tensions, ensuring the well-being of communities (Vujcic, Tomicevic-Dubljevic, Zivojinovic, & Toskovic, 2019) and guaranteeing environmental justice for all walks of life.

REFERENCES

Cioci, S., Eckelman, M., & Onnis-Hayden, A. (2019). The strategic green infrastructure: quantification of environmental and economic benefit for the city of Somerville. UpLand, Giornal of Urban Planning, Landscape and Environmental Design. https://doi.org/10.13140/RG.2.2.20343.37284.

Henein, K., & Merriam, G. (1990). The elements of connectivity where corridor quality is variable. Landscape Ecology, 4(2–3), 157–170. https://doi.org/10.1007/BF00132858

La Greca, P., & Carta, M. (2017). Cambiamenti dell'urbanistica: responsabilità e strumenti al servizio del paese. Donzelli.

Salzano, E. (2012). Fondamenti di urbanistica: la storia di urbanistica. Retrieved from http://www.laterza.it/index.php?option=com_laterza&Itemid=97&task=schedalibro&isbn=9788842071525.

Sykes, O., & Brown, J. (2015). Capitali europee della cultura e rigenerazione urbana: prospective urbanistiche da Liverpool. Urbanistica, 67(155), 76–82.

US EPA. (2014). The Economic Benefits of Green Infrastructure A Case Study of Lancaster, PA About the Green Infrastructure Technical Assistance Program. Retrieved from http://water.epa.gov/infrastructure/greeninfrastructure/gi_support.cfm.

Vujcic, M., Tomicevic-Dubljevic, J., Zivojinovic, I., & Toskovic, O. (2019). Connection between urban green areas and visitors' physical and mental well-being. Urban Forestry and Urban Greening, 40, 335–343. https://doi.org/10.1016/j.ufug.2018.04.004.

Pedestrians, Urban Spaces and Health – Tira, Pezzagno & Richiedei (eds)
© 2020 Taylor & Francis Group, London, ISBN 978-0-367-46171-3

Boston healthy city: The harborwalk experience

L. Kappler
DICEA, "Sapienza" University of Rome, Italy

ABSTRACT: Boston is a city of water, built on a field artificially created with landfill operations, which have buried ponds and coastal marshes, widening the originary area from 3 sq km to the current 125. Among the projects aimed at preserving the usability of the waterfront, lapped and eroded by the ocean, is the 43 miles linear path known as Harborwalk. It is expected to underline how the answers to challenges as the water issue and urban cohesion, come from overall assessments of the administration that find in the Harborwalk a synthetic one event for the city's health and citizens' involvement.

1 BOSTON CITY OF WATER

Walking through Boston, the perception of the urban context does not allow us to imagine that once most of the land that is the basis of the oldest parts of present-day city did not exist. The landscape was constantly modified by water, a predominant and changeable element. The Puritan colonists in 1630 settled in what was called by the Native Americans "Shawmut", a small peninsula of 3 sq. Km, connected to the mainland by a narrow neck that was submerged during the high tide. The guarantee of a reliable supply of spring water and the strategic position for maritime trade had made the area the best conquest to rise a florid city. Among the main causes of the need for new land there were the desire to strengthen the port to compete with other cities, the growing attention to the protection of public health and the need to have parks and attractive neighborhoods to encourage the inhabitants to stay. The landfill operation began in 1807, at Mill Pond, the current Bulfinch Triangle, with material from Beacon Hill. Similar interventions followed until the end of the 19th century, giving rise to the current Back Bay, Bay Village, South End, Fenway, Chinatown, Leather District, North End, Downtown, Fort Point Channel and parts of East Boston (Kappler, 2018). If in the historical areas the raising of the water level is necessary to preserve the existing heritage, to maintain adequate compression on the deep foundations of wooden poles that support the buildings, on the contrary, in the areas lapped by the ocean it has always aimed at lowering and containment to limit erosive activity.

2 MAYOR FLYNN'S VISION

Originally envisioned by then-Mayor Raymond Flynn in 1984 in his Harborpark Framework, the Boston Harborwalk arose to protect public access to the waterfront and as a buffer zone, when the city started the redevelopment of former industrial wharves. Boston Harbor led the country into the mercantile period of the 18th century and helped finance the industrial revolution of the 19th century. Boston's economic competitiveness, however, waned during The Great Depression of the 1930s. A shift in patterns of trade and the location of manufacturing activities after the War resulted in decaying port facilities and abandoned warehouses and factories. Boston's economic revival, from the early 1960's was initially spurred by the growth of non-Harbor related service activities. In the following decade, the attractiveness of the waterfront has drawn development interests to the Harbor's edge, as a vital source of growth for the economy. Harborpark was meant to be a framework for discussing the ordering of this growth. It was designed to guarantee public access to the environment along the Boston Harbor, while encouraging balanced growth along the entire waterfront. Each pier and wharf retained its own identity, yet it was imagined to integrate each area into an uninterrupted and free from barriers walkway, the Harborwalk, from Charlestown to South Boston to combine a diversity of uses: maritime and commercial activity, new housing for every income, arts facilities, open space amenities, existing parks (Esplanade) and paths (Freedom Trial). This planning process should have rekindled the spirit of community in the place of Boston's origins. "Harborpark Phase One", in line with the Chapter 91[1] indications, proposed the Harborwalk as a continuous 7 mile public waterfront path designed with ribbons of green, connecting the wharves and linking waterfront cultural and recreational activities outlined in the plan.

3   COASTAL FLOODING RISKS

Sea level rise puts filled tidelands and other low-lying areas at growing risk. Once these risks became clear, the insufficiency of the extension, of the design and security measures of the Flynn's Harborwalk was evident. Boston's sea level could rise from 2013 levels by nine inches by the 2030s and 40 inches by the 2070s. With sea levels 40 inches higher, flooding across most of South Boston, with the exception of the historic residential neighborhood, will occur at least once a year. The flood pathways along Fort Point Channel and Seaport Boulevard can be addressed with near-term strategies along the shoreline. As sea levels rise, these flood pathways merge with more widespread flooding in South Boston and, later with no action, flooding could reach South End and connect with flood pathways from the Charles River and Moakley Park.

4   IMAGINE BOSTON 2030 AND CLIMATE READY BOSTON

With the release of the Imagine Boston 2030 and Climate Ready Boston plans, the City has formally committed itself to supporting the proactive implementation of a strong waterfront vision that balances development, open space and the maritime industry, preparing the city for the next environmental transformations. The 2016 Climate Ready Boston report assessed Boston's vulnerabilities to climate change, and developed city-wide strategies for reducing vulnerability to sea-level rise, coastal flooding, more extreme heat and more intense precipitation. Climate Ready Boston implements the Greenovate Boston 2014 Climate Action Plan Update strategy of integrating climate preparedness into city planning, review and regulation. Similarly, one of the goals of Imagine Boston 2030, the first city-wide comprehensive plan in 50 years, is to "promote a healthy environment and adapt to climate change." Coastal resilience solutions include the Harborwalk, elevated waterfront open spaces, reinforced structures and piers, flood walls, dunes and a living shoreline. Achieving these changings will require public investments, private action and support through regulatory change. By the 2030s, these actions will protect residents, businesses, drainage and combined sewer systems, critical highway, transit infrastructure, first responder facilities, healthcare facilities and redevelopment areas. Investment in coastal resilience solutions will prevent billions of dollars in physical damages and displacement costs.

---

1. Chapter 91 is a Massachusetts General Law from 1866. Based on the ordinances of Colonia Bay Massachusetts (1640) it promotes public use of waterways in the Commonwealth. Today it is the oldest law of its kind in the nation.

4.1 *Coastal resilience solutions*

Nearly 700 Bostonians participated in meetings and interviews to share their priorities for coastal resilience. The measures proposed included the enhancement of the Harborwalk to improve connections to the whole waterfront in the city, site amenities such as seating, steps and furnishings, elevated waterfront parks and natural wetland buffers. They would protect up to the 1-percent annual chance flood level with 40 inches of sea level rise (2070s), plus 1 foot of freeboard. This adaptability could ensure effectiveness for about 70 years.

5 THE HARBORWALK TODAY

Incorporating the addresses of the plans and coastal resilience solutions, the Harborwalk has become a linear park 43 miles along the coast of Boston (Figure 1), that connects the Boston neighborhoods on the seafront between them and the port of Boston. Nearing completion, it extends from the Neponset River in the lower Dorchester to Constitution Beach in East Boston via Charlestown, North End, Downtown, Fan Pier (Seaport), Fort Point Channel, South Boston and Dorchester. It is made recognizable by the choice of a clear stone floor. Part of the wealth of Harborwalk is its variety, which reflects the various activities and urban plots of the adjacent land. At some points, it extends to the maritime industrial areas, allowing visitors to observe the port operations at close quarters. In others, the Harborwalkers can enjoy a swim, go fishing, visit over forty parks and a dozen museums, or eat. The Harborwalk connects not only to the piers and new buildings, but to 9 public beaches, to numerous foot-paths and parks, including the Emerald Necklace, Charles River Esplanade, the Rose Kennedy Greenway, the Freedom Trail, the South Bay Trail and the East Boston Greenway. It is a non-motorized route with the exception of wheelchair accessibility. Bicycles are welcome from the Neponset River Greenway to Castle Island State Park. The Harborwalk includes green lawns, seating areas, public restrooms, fishing docks and observation decks. It presents 365 individual properties with a variegated rose of concessionary status and owners which include the city of Boston, the Commonwealth of Massachusetts, the federal government and private landowners.

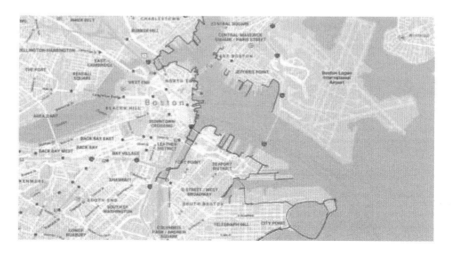

Figure 1.   The core of the harborwalk benefits map, accessed August 2019.

6 THE HARBORWALK GOALS

The Harborwalk pursues three fundamental objectives: healthy people, connecting people, promoting a healthy use of the coastline that encourages pedestrian movements, but also establishing a strong sense of community and protecting the territory with the establishment of associations such as Boston Harbor Now, committed to protecting it; healthy economy, supporting the balanced development of activities in the waterfront in coherence with the new active and innovative port; healthy environment, to keep the waterfront healthy, safe and accessible.

6.1 *Healthy people*

The involvement of the citizens is represented by Boston Harbor Now, the lead nonprofit focused on Boston Harbor, its waterfront and islands. Its work is focused on increasing the quality, destination value and overall awareness of Boston Harbor and its importance to the environmental, social, and economic health of our city and region.

Boston Harbor Now works in partnership with others in the public, for profit, and non-profit sectors to plan, advocate and activate a thriving waterfront, harbor, and Boston Harbor Islands National and State Park. Boston Harbor Now's comprehensive approach to the harbor prioritizes equitable public access through the Harborwalk, parks and open spaces; infrastructure and ferries; maritime industry and other marine-related economic drivers; mixed-use and mixed income climate-resilient waterfront development; planning of recreational and educational uses; maintenance of the clean harbor. The nonprofit provides free or low-cost public programming from concerts and theater to living history and educational field trips.

6.2 *Healthy economy*

Boston Harbor Now is partnering with the Massachusetts Department of Transportation to help bring Boston Harbor's water transportation system to scale and link it to land-based transit options. Boston Harbor ferries and water taxis would help ease Greater Boston's traffic problems, spur transit-related development, and increase recreational access to the Boston Harbor Islands. Moreover, in early 2018, Boston Harbor Now convened experts to discuss the working waterfront issues and explore solutions employed by other national and international port cities. This debate led to "The 2018 report Innovation in Boston's Working Port: Planning a 21s t -Century Harbor". It set the stage for an overdue discussion of the working harbor and the significance of its contribution to the city and the region's economy. The report focuses on developing recommendations for the Boston's working waterfront and on four emerging themes: growth, flexibility, synergy, and change. Concerning growth, under-used or vacant areas offer multiple opportunities for improvement. To remain competitive, the city should go beyond preserving waterfront parcels through designated Port Area Master Plans and begin to invest in its competitive advantages for new innovative, cutting edge ideas and business concepts. The elevation of the public presence and the creation of a strategic business alliance among maritime industry leaders could lead to understand future maritime trends and capitalize on key opportunities. That is why recently Massport has created the Port Industry Alliance, a coalition of Boston's maritime leaders to improve strategic collaborations, adapting to external changes as e-commerce and evolving regulatory requirements as cleaner fuels.

6.3 *Healthy environment*

The Friends of the Boston Harborwalk is a group of volunteers formed in 2014 to create awareness, promote public enjoyment and foster local stewardship of the Harborwalk. The group meets monthly to plan and coordinate its priorities. It hosts informative tours highlighting different path's segments and develops interpretive signs. Through the Harbor-walk Clean-up Team the group organizes waterfront clean-up events and reports maintenance concerns to help ensure cleanliness and safety.

7 RECENT INITIATIVES

In February 2018 it was announced that the Massachusetts Department of Environmental Protection (MassDEP) would have directed mitigation funds from The Fallon Company 21-acre Fan Pier development (2007), which is surrounded by the Harborwalk on three sides as part of their Chapter 91 license agreement with MassDEP, to the creation of the Boston Harbor Now Harborwalk Benefits Map and public waterfront amenities database. The user-friendly map and database would help visitors to navigate the waterfront and understand the inventory of public amenities along the path. Boston Harbor Now conceptualized the Harbor-walk Benefits Map after identifying a lack of knowledge among properties and waterfront stakeholders regarding opportunities along the Harbor. There was no map about these spaces, making it difficult to understand what is publicly available. The Harborwalk Map is available online from spring 2018 for free. Boston Harbor Now continues to convene with property managers and stakeholders along the waterfront to discuss ways to ensure residents feel welcome visiting the Harbor through the Harborwalk.

REFERENCES

City of Boston. (1984). Harborpark. BRA.
Cioci S., Kappler L., Mattogno C., (2018). Boston: una foresta di pali sommersi. Urbanistica Informazioni 278. INU edizioni. 8–11.

Pedestrians, Urban Spaces and Health – Tira, Pezzagno & Richiedei (eds)
© 2020 Taylor & Francis Group, London, ISBN 978-0-367-46171-3

The node-place model to improve walkability in railway station catchment areas to promote healthy city environments. An application to the municipality of Cercola (NA)

G. Carpentieri, C. Guida & L. Faga
Department of Civil. Building and Environmental Engineering. University of Naples Federico II, Italy

ABSTRACT: This contribution presents experimentation of research on urban accessibility applying to professional consultancy work, done by the Department of Civil, Architectural and Environmental Engineering of the University of Naples Federico II, for the city of Cercola (NA) to support the edit of the City Master Plan. One of the more complex challenges in urban planning is the poor connection between the transport system and the land-use pattern, which encouraged the use of the private car as the first means of transport, polluting and overcrowding urban areas. In this study, we propose the application of Node-Place method at the station level to improve the urban quality of life, safety and economic condition. The city of Cercola is in the periurban area of Naples, through the application of Node and Place model at the local scale, we defined the potential solutions to make the city more accessible and the definition of sustainable mobility solutions.

1   INTRODUCTION

The increasing share of urban citizens and the sustainability paradigm declined in its social, economic and evironmental features represent the new challenges (and opportunities) that cities need to face during next decades. Hence, it is crucial for urban and regional systems to develop new tools, approaches and guidelines for the analysis, quantification and solution of urban issues (Leone, Gobattoni & Pelorosso, 2014; Papa. Angiello & Carpentieri, 2018). In fact, urban planning experciences of last century proved that a lack of coordination between planning practice of land use and transport components can lead to adverse environmental and socioeconomic effects on mobility and activity systems. The reduction of urban traffic congestion, as well as the pollution caused by an excessive mobility through private transport mode, and the transformation of built environment in order to make cities more accessible, are some of the aims arisen in Agenda 2030 goals. In scientific literature field, different methodologies to integrate transport and land-use systems in territorial planning practices have been studied, but they hardly ever found application in real practice (Bonotti et al., 2015). One of the most known methods is the Node-Place model, developed by Bertolini in 1998, whose aim is to classify railway stations catchment areas in order to suggest useful guide lines for administrations to improve these areas, both as "places" of activities, relationships and services, and as "nodes" of the multimodal transport network. Considering Bertolini first research, railway stations could be in a dependence condition if both the node and place indexes are very low, since they count on bigger stations for their transport feature as for their activities and services. Railway station could be stressed if their indexes are both high, compared to the network values, and interventions aimed at reducing passengers, employees and dwellers flows should be introduced. While, if the two indexes are not balanced different interventions are needed, to improve their role in the transport network or to increase activities and services to make stations more livable places, From its first application to Netherlands railway stations, the model has been enhanced and applied in different contexts (Switzerland, Toronto, Lisbon, Benjin, etc.) and it comes out that the original structure of the model is not applicable univocally to the different territorial contexts. In particular, the number and type of indicators must be changed to consider the different physical-functional characteristics of the case study and the availability of data for the calculation of the indicators (Carpentieri & Papa, 2018).

2 OBJECTIVE OF THE PAPER

In order to fill the gap between literature and real practices, the aim of our contribution is to present a methodology to be implemented in drafting City Master Plans, in order to improve walkability and accessibility within railway station catchment areas, characterised by a lower-density for the population and the activities and a mono-functional land use destination.

3 METHODOLOGICAL APPROACH

Considering the original application of the Node-Place model, several adjustments were needed to consider the different context for its extention and main characteristics. First of all, from Bertolini and other scholars' and experts' evaluations, a set of significative variables to model both the node and the place features of the railway stations cathment area was seleceted. Then, a GIS-based procedure was designed to compute the values of each Node and Place indicator and to develop the spatial analysis (Campagna. 2014).

4 DESCRIPTION

The methodolody is based on the evaluation of two syntetic indexes. the Node and the Place, and on their relationship, since they quantitatively define the characteristics of railway stations catchment areas, both as nodes of the transport network graph and as living places in the urban environment. By a systematic review of literature, a set of indicators for Node and Place evaluations were selected. as reported in Table 1.

Once the dataset is ready, a GIS geodatabase needs to be created to better organise multiple sources of spatial and alphanumeric data. Then, the values of ten indicators are calculated for each station node and CAs (place) of the network. Moreover, all indicators are standardised to have a minimum value of 0 and a maximum of 1 (Reusser et al., 2008). The synthetic indices of Node and Place are computed as the averages of all standardised values of the same indicators category, and represented in a chart.

4.1 *Innovative aspects*

The most innovative aspect of the proposed methodology concerns that the selected indicators of Place have a direct impact on urban planning so that they could be easily controlled by local government tools, such a City Master Plan. This can be useful both fot the knowledge phase, as well as for the decision and monitoring phases.

4.2 *Technical developments*

The methodology was applied for the Cercola Municipality, to draft the City Master Plan, oriented to decrease the use of private cars. It is located in Campania Region, on the slopes of Vesuvius volcano, and about 19.000 people reside in the municipality.

The city is served by the Napoli – Ottaviano – Sarno line, managed by EAV transport company, with one railway station located in Via Antonio Gramsci (in the South-West area of Cercola). Figure I describes the territorial context of the railway station and the

Table 1.   The set of place and node indicators selected for the methodology.

Index	ID	Indicator	Measurement	Data source
Place	P1	Station Area	Extension of station catchment area. within 800 m of walkable distance	OSM
	P2	Employees Density	Density of employees within station CA	ISTAT
	P3	Population Density	Population density within station CA	ISTAT
	P4	Land Use	Distribution of high human density activities within station CA	Google Maps data
	P5	Residents mobility	Average number of movements for the city dwellers	ISTAT
	P6	Employees Mobility	Average number of movements for work purpose	ISTAT
Node	N1	Intermodality	Number of nodes for different transport mode. within station CA	Google Maps data
	N2	Average frequency tax	Flows of passengers. considering their distribution in a working day	Google Maps data
	N3	20 minutes of travel	Number of nodes reachble within 20 minutes of travel	OSM and Google data
	N4	Time from CBD	Travle time to get to the nearest Business District	OSM and Google data

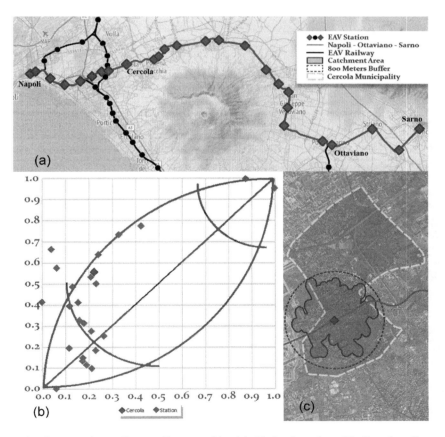

Figure 1.   Apoli – Ottaviano - Sarno railway corridor (a); Node-place chart (b); Cercola railway station CA (c).

Table 2. Place and node indexes.

Stations	P1	P2	P3	P4	P5	P6	N1	N2	N3	N4
Porta Nolana	0.98	0.92	1.00	0.64	0.74	1.00	1.00	0.78	0.90	0.99
Garibaldi	0.26	1.00	0.87	0.62	1.00	1.00	1.00	0.91	0.90	1.00
Via Gianturco	0.64	0.27	0.07	0.00	0.27	0.07	0.20	0.87	0.72	0.96
San Giovanni a T.	0.50	0.25	0.26	0.52	0.49	0.11	0.20	0.78	0.34	0.93
Barra	0.91	0.07	0.37	0.82	0.51	0.21	0.40	0.81	1.00	0.89
Ponticelli	0.64	0.06	0.51	0.95	0.32	0.03	0.20	1.00	0.93	0.84
Vesuvio De Meis	0.43	0.01	0.28	1.00	0.36	0.06	0.40	0.77	0.72	0.78
Cercola	0.94	0.04	0.17	0.73	0.17	0.03	0.20	0.53	0.90	0.78
Pollena Trocchia	0.05	0.02	0.07	0.86	0.12	0.04	0.20	0.29	0.76	0.72
Guindazzi	0.46	0.00	0.02	0.92	0.24	0.04	0.20	0.57	0.72	0.71
Madonna Dell'A.	0.00	0.03	0.10	0.83	0.35	0.09	0.40	0.84	0.55	0.69
Sant'Anastasia	0.33	0.12	0.27	0.79	0.42	0.09	0.40	0.99	0.10	0.85
Villa Augustea	0.67	0.01	0.09	0.91	0.10	0.04	0.20	0.24	0.66	0.59
Somma	0.66	0.03	0.11	0.89	0.21	0.00	0.08	0.78	0.05	1.11
Rione Trieste	0.58	0.00	0.00	0.65	0.36	0.04	0.20	0.86	0.34	0.51
Ottaviano	1.00	0.05	0.14	0.73	0.19	0.00	0.00	0.44	0.38	0.43
San Leonardo	0.51	0.01	0.02	0.73	0.38	0.00	0.00	0.90	0.00	0.39
San Giuseppe	0.85	0.12	0.17	0.56	0.18	0.00	0.00	0.43	0.34	0.36
Casilli	0.75	0.03	0.04	0.59	0.00	0.00	0.00	0.00	0.34	0.33
Terzigno	0.83	0.04	0.13	0.68	0.21	0.04	0.20	0.51	0.03	0.29
Flocco	0.64	0.01	0.10	0.88	0.41	0.00	0.00	0.97	0.38	0.20
Poggiomarino	0.73	0.05	0.22	0.83	0.30	0.11	0.20	0.71	0.38	0.18
Striano	0.70	0.03	0.07	0.67	0.40	0.00	0.00	0.95	0.07	0.08
San Valentino T.	0.60	0.02	0.09	0.82	0.28	0.09	0.40	0.67	0.59	0.00
Sarno	0.67	0.10	0.13	0.58	0.18	0.13	0.60	0.58	0.38	0.40

results of the model. Table 2 shows the model results for each station of the railway corridor considered.

This methodology could have potential use not only in the main urban centers, but they could be useful to the development of suburban areas. which suffers from poor connectivity in terms of public transport and access to main services localized in the principal urban areas.

5 RESULT OF THE RESEACH

The application of the methodology to the Cercola railway station lead to some significant results. From the Place-Node chart it emerges that the station has a "Place-deficit", probably due to a broadly monofunctional activity system. In order to equilibrate the station, interventions related to the accessibility and walkability within the station catchment area are needed. The methodology does represent a strong tool to support public administration in order to better guide its decisions in defining the local developments plans and allocating economic resources.

6 CONCLUSIONS

The Place-Node model presented in this contribution was applied to support public administrators' decision process and it lead to significant results, since it emerged that a place-based approach is necessary to improve the livability of the station CA. Although the model is applied just for the station CA, when drafting a City Master Plan, it could be the engine for a holistic vision of the urban system, more oriented to the livability and walkability of its environments. Moreover, since the station is part of a more extended network, interventions on the single area could become the driving force for other transformations in the whole corridor territory, which, in this case, goes from Naples to the Salerno Province. Considering the new urban chellenges, territory planners have to face them as opportunities of growth, rather than weaknesses. The integration between transport system and land-use pattern is the potential key to address future urban challenges, according to the sustainability paradigm.

NOTES

In light of the team work at the basis of this contribution, paragraphs 1, 2, 3, 4, 5 and 6 were made by Gerardo Carpentieri and Carmen Guida; subparagraph 4.2 and image and data elaborations were made by Luigi Faga.

REFERENCES

Bertolini. L., (1999). Spatial Development Patterns and Pubblic Transport: The Application of an Analytical Model in the Netherlands. Planning Pratice & Research. 14(2),199-210. doi: https://doi.org/10.1080/02697459915724.

Bonotti, R., Rossetti, S., Tiboni, M., & Tira, M. (2015). Analysing Space-Time Accessibility Towards the Implementation of the Light Rail System: The Case Study of Brescia. Planning Practice & Research, 30(4),424-442. https://doi.org/10.1080/02697459.2015.1028254.

Campagna. M., (2014). Geodesign from theory to practice: From metaplanning to 2nd generation of planning support systems. Tema. Journal of Land Use. Mobility and Environment.

Carpentieri, G., & Papa, R. (2018). Classifying railway station catchment areas. An application of node-place model to the Campania region.

Leone. A., Gobattoni. F., & Pelorosso. R. (2014). Sustainability And Planning. Thinking and Acting According to Thermodinamics Laws. Tema. Journal of Land Use. Mobility and Environment.

Papa. E., Carpentieri. G., & Angiello. G. (2018). A TOD Classification of Metro Stations: An Application in Naples. In Smart Planning: Sustainability and Mobility in the Age of Change (pp. 285–300). Springer. Cham.

Reusser. D. E. Loukopoulos. P. Stauffacher. M. & Scholz. R. W. (2008). Classifying railway stations for sustainable transitions–balancing node and place functions. Journal of transport geography. 16(3). 191–202.

Pedestrians, Urban Spaces and Health – Tira, Pezzagno & Richiedei (eds)
© 2020 Taylor & Francis Group, London, ISBN 978-0-367-46171-3

HSR stations' urban redevelopments as an impulse for pedestrian mobility. An evaluation model for a comparative perspective

M. Carra & P. Ventura
University of Parma, Italy

ABSTRACT: Return to more compact cities, own of the pedestrian movement, a transit-oriented development (TOD) all must be promoted as a model of urban design, in particular for the areas around the railway stations. The design of these areas requires a balance between the intermodal characteristics of the node and the characteristics of intensity and proximity of functions in the place, according to a relationship of micro and macro-accessibility. Through an integration of the different node-place models, the present contribution focuses on the multi-criteria analysis of urban projects and dynamics generated in the area of HSR stations, both in the transformation of the existing ones and in the context of the construction *ex-novo*. The analysis aims to evaluate stations according to aspects of urban planning and design: location, intermodality, land use and accessibility. This paper is part of preliminary research of a PhD at Parma Doctorate School.

1 INTRODUCTION

At the global scale, urban development phenomena undergo a constant increase in urbanization (Burdett & Rode, 2018)(Angel, et al., 2016), the reduction in the density of resident population (through extensions, infill and leapfrog), an expansion that erodes the agricultural territory melding together pre-existing urban settlements through capture phenomena. Consequently, priority to renewal, regeneration, and adaptation of appropriate urban areas becomes primary (Roberts, Sykes, & Granger, 2017)(Hall, 2014). Interventions based on the city transformation require a seamless integration between urban design and urban planning, which must pursue the following objectives: urban expansion control, containment of land consumption, equipping the city with buildings and public spaces of high quality, protection of agricultural uses in built-up areas, in fringe open spaces and in urban extensions. The identification of areas suitable for mutation processes is closely related to the quality and quantity of flows of goods and people, determined by the proximity of valuable types of functions and mobility infrastructures. In summary, public transportation nodes are potential factors of location, size and physical configuration that can attract higher-level functions.

Railway stations, after at least 20 or more years of neglect, are undergoing new attention by planners, promoters and cities to revitalise urban cores. Since the 1970s, high-speed trains have led to an increase in accessibility with macro effects, at the local scale, and micro, at the urban scale. At the macro scale one can identify four main effects: a) the substantial acceleration of the space-time contraction phenomena (Spiekermann & Wegener, 1994); b) the competitiveness with the airport in the case of medium-distance journeys (400-600 km) (Vickerman, 2015); c) the largest deficit between cities that are part of the network and those that are not, with reverberation on economic growth (van den Berg & Pol, 1998); d) greater competitiveness in urban areas with direct effects on the liveability and quality of the urban environment.

The railways' companies operated several urban projects of new stations (Thorne, 2001) and the revitalization of existing stations to optimise the effects generated at the micro-urban scale and to attract more train-customers. These projects recover the previous 'monument symbol of technological conquest'. Some of these realizations are part of ambitious urban regeneration programs with synergic effects in and between nearby and distant cities (Garmendia, Ribalaygua, & Ureña, 2012).

According to the TOD *rail-centric* model, the complexity of the urban design of these areas depends on a balance at least between the following three factors: the intermodal characteristics of the station (node); the characteristics of land use (place), the pedestrian accessibility and practicability to services close to the transport hub (Calthorpe, 1993).

Accordingly, several studies presented application models for assessing the potential of these areas and subsequent urban regeneration programs, starting from the node-place model (Bertolini & Spit, 1998) and subsequent implementations (Peek, Bertolini, & De Jonge, 2006). These models are a basic reference of the present research for multi-criteria analysis.

2   OBJECTIVE OF THE PAPER

The general aim of this paper is to discuss the relationship between the insertion of HST stations and urban development and regeneration program. Several models have been analyzed and multiple configurations have been adopted. A first objective focuses on identifying factors and interactions that influence the spatial performance of areas annexed to these stations. The identified spatial performance is then applied to a model. It is a tool that can easily understand balances/imbalances between the identified elements, a tool useful in urban planning and design of future regeneration projects in the existing city. A second objective concerns the application of the model to case studies.

3   METHODOLOGICAL APPROACH

The approach focuses on the two station features, as 'node' of the transport network and as urban 'place' (Bertolini & Spit, 1998). To highlight the spatial performance that interacts in the area, different developments and operational or theoretical implementations of the model have been analyzed and compared (Peek, Bertolini, & De Jonge, 2006)(Meijers, Drenth, & Jansen, 2002)(Papa, 2006). Additional models focus on the interaction between potential factors, the classification of areas and definition of strategies (van Hagen & Bruyn, 2002)(Vaessens, 2004). The set of models allowed us to outline recurring factors. Starting from the *Vlindermodel*, developed by the Metropool Association and applied to Rotterdam (Vereniging Deltametropool, 2013), the present research has outlined the implementation of the multi-criteria analysis of qualitative/quantitative elements. The methodological approach is here applied to the case of three HST station area in medium-size cities.

4   DESCRIPTION

4.1   *Innovative aspects*

Case studies processing is calculated as an evaluation of the walkability environment. Through an urban and transport survey, the information has been collected and then exported in a specific diagram, able to illustrate its qualities and quantities. It allows to highlight the peculiarities or critical issues as it is and of future urban design choices, but also to observe and to facilitate the understanding of possible transformations.

4.2   *Research and technical developments*

The model is configured as a radial graph (Figure 1) which sets the value of the node in the left and the value of the place in the right. Both values are composed of three sub-parameters:

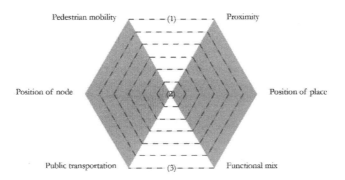

Figure 1. Model of multi-criteria analysis.

characteristics of the position, and micro and macro-accessibility. The symmetry of the data determines a complementarity between the node and place elements.

The *pedestrian mobility* is measured starting from the HST station through a ratio between the buffer and the flexible Service Area (Zazzi, Ventura, Caselli, & Carra, 2018) of a 15-minute walk. The value of the *proximity*, i.e. the ratio between the total surface of active commercial and service functions and the pedestrian isochrone, hints the degree of functions easily reachable by walking. The level of *public transport*, is given by the sum of the connectivity, a value depending on the availability of specific modes of public transport (HST, intercity, regional trains, metro, tram, bike sharing), and the value of accessibility, measured as a product between the number of directions and their types of connections. Values related to vehicular mobility are not taken into account. The presence of a *functional mix* is measured through a value among the functions in the pedestrian area: min, max, med, sum. Lastly, the *position of the node* is interpreted by the sum of two judgment values: the first is the station' area location (internal, nearby, peripheral or external to the urban core), which expresses the degree of integration with the city; the second concerns the position of the railway and expresses the degree of interference on the area (embankment, suspended, grade-level, underground). This factor is then compared to the value of the *position of the place* given by the sum of city size class and the place location' (both judgment values) and the product of the percentage of the urbanized area, that provides indications to the presence of available areas. To compare data, values are normalized according to a ranging from a minimum of 0 to a maximum of 1.

5   RESULTS OF THE RESEARCH

The results allow the reading of four balances: between the node and the place; between the sub-parameters placed on the same level as the graph.

Specifically, the degree of synergy between the attractiveness and the use of pedestrian (1) and infrastructural (2) accessibility to the area functions; the degree of area transformability, dependent on the presence of available areas, the presence/absence of interference or conflicts, and the synergy between more or less advantageous location factors (3). The model allows comparing multiple nodes of HST stations, in two different moments: before (T0) and after the regeneration project (T1)(Figure 2).

In the time T0 the results show a general balance between the value of node-place location', according to rather reduced values in the cases of Avignon (France) and Reggio Emilia (Italy), where the new station interferes with the agricultural context outside the existing built area (PN: 0.25; 0.25. PP: 0.23; 0.20). In Valladolid (Spain), the high speed is part of the existing station, located in a central position within the urban core (PN: 0.55; PP: 0.66). This case also shows a good balance between the other two values with a higher variation between the

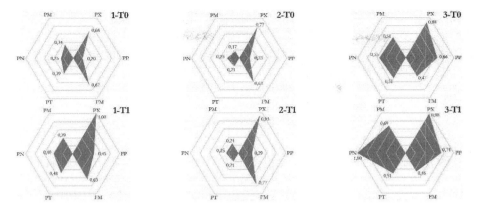

Figure 2.   Model's application to case studies of Avignon (1), Reggio Emilia (2), Valladolid (3).

proximity (0.88) and the pedestrian isochrone (0.51). Avignon and Reggio Emilia case studies instead show strongly unbalanced values: medium-high in proximity (0.77; 0.68), due to a low value of the pedestrian isochrone; medium in the functional mix (0.62; 0.67), characterized by a commercial and industrial function with the presence of residential buildings integrated to the factories. In the time T1 the results show a general increase of the values and a tendency to greater balance. In Avignon case study the redevelopment and expansion project generates a new district of the city by changing the position of the station from external to peripheral. Consequently, the increasing balance between the value of node-place location' remains stable (0.05). A balance' improvement concerns public transport and functional mix, determined by an increase (+0.1) of the connectivity of the first (through a regional train with the central station and with the planned HSR with Montpellier) and the decrease in the second (-0.04) due to the prevalent inclusion of residential and tertiary functions to the detriment of industry. The highest imbalance is between proximity and pedestrian mobility (+0.27), due to the interference of the railway and, beyond it, the presence of the Durance river. Valladolid case study shows positive values between public transport and functional mix (0.05) and between pedestrian mobility (+ 0.18) and proximity (+0.10), thanks to the drop of the line and the consequent higher permeability of the area. The removal of the interference affects the value of node location' (+0.45). In contrast to Avignon and Valladolid, the transformation project of Reggio Emilia maintains large imbalances. The cause is due: to the lack of intermodality of public transport and node location (+0.00); to the failed measures to increase pedestrian accessibility (+0.04) beyond the limit of the embankment line. The project determines an improvement in the functional mix (+0.15) and proximity (+0.14). The data highlight that the site can be reached mostly from the motorized vehicle.

6   CONCLUSIONS

The model can adequately collect a series of spatial factors that define the quality and quantity of spaces attached to the HST stations. The graphic rendering is clear and easily comparable between the three analyzed cases and between the subsequent regeneration projects. Moreover, it allows having more levels of relationships between the considered elements. The addition of a pedestrian evaluation of the area is fundamental for a TOD. Indeed, the perimeter of the actual usable area allows identifying three different areas of development, depending on the distance: primary within 10 minutes; complementary between 10 and 15; semi-complementary beyond 15 minutes, with functions only partly dependent on high-speed.

6.1 *Barriers and drivers*

The analysis will be extended to a wide range of case studies to determine a series of records. These records can be classified and used appropriately for possible developments in the pedestrian areas reachable from the HST stations. These classifications can constitute a basic element to set up specific guidelines and possible resolutions to be adopted. The research intends to apply this address to proceed with a series of project simulations for the Brescia case study, for which no interventions have been planned yet.

The comparison between typologies could also allow assessing whether there are some interactions between one or more stations, for which the improvement of one can inspire the improvement of another.

REFERENCES

Angel, S., Blei, A., Lamson-Hall, P., Galarza Sanchez, N., Gopalan, P., . . . Shingade, S. (2016). Available at: Atlas of Urban Expansion: http://www.atlasofurbanexpansion.org (accessed 14.05.2018).

Bertolini, L., & Spit, T. (1998). Cities on rails: the redevelopment of railway station areas. London: E & FN Spon.

Burdett, R., & Rode, P. (Edited by). (2018). Shaping cities in an urban age. London, New York: Phaidon.

Calthorpe, P. (1993). The Next American Metropolis: Ecology, Community, and the American Dream. Princeton: Princeton Architectural Press.

Garmendia, M., Ribalaygua, C., & Ureña, J. M. (2012). High speed rail: implication for cities. Cities, 29 (2), pp. 26–31.

Hall, P. (2014). Good cities, better lives: How Europe discovered the lost art of urbanism. London, New York: Routledge.

Meijers, E. J., Drenth, D., & Jansen, A., A. (2002). Mobiliteit en Beleid. In F. Bruinsma, J. van Dijk, & C. Gorter (A cura di), Knooppunten en mobiliteit (pp. 109–121). Assen: Koninklijke Van Gorcum.

Papa, E. (2006). Trasformazione urbana e sistemi di trasporto su ferro: da un paradigma interpretativo ad un caso di studio. Università degli Studi di Napoli Federico II, Di.Pi.S.T. Napoli.

Peek, G.-J., Bertolini, L., & De Jonge, H. (2006). Gaining insight in the development potential of station areas: A decade of node-place modelling in The Netherlands. Planning Practice & Research, 21(4), pp. 443–462.

Roberts, P., Sykes, H., & Granger, R. (A cura di). (2017). Urban Regeneration (II)ed. London: SAGE Publications.

Rossetti, S., Tiboni, M., Vetturi, D., & Enrique, J. C. (2015). Pedestrian mobility and accessibility planning: some remarks towards the implementation of travel time maps. Journal City Safety Energy(1), pp.67–78.

Spiekermann, K., & Wegener, M. (1994). The shrinking continent: new time-space maps of Europe. Environment and Planning B, 21, pp. 653–673.

Thorne, M. (2001). Modern Trains and Splendid Station: architecture, design, and rail travel for the twenty-first century. London: Merrell.

Vaessens, B. (2004). Synergie op stationslocaties. Bijdrage aan het Colloquium Vervoersplanologisch Speurwerk 2005, 24–25.11.2004, (pp.1–20). Antwerpen.

van den Berg, L., & Pol, P. (1998). The European High-Speed Train and Urban Development: experiences in fourteen European urban region. Aldershot: Ashgate.

van Hagen, M., & Bruyn, M. (2002). Typisch NS. Elk station zijn eigen rol. Colloquium Vervoersplanologisch Speurwerk 2002: De kunst van het verleiden, 28 en 29 november 2002, Delft, (pp.1–20). Delft.

Ventura, P. (2004). Città e stazione ferroviaria. Firenze: Firenze University Press.

Vereniging Deltametropool. (2013). Knooppunten in de Stadsregio Rotterdam. Rotterdam: Deltametropool.

Vickerman, R. (2015). High-speed rail and regional development: the case of intermediate stations. Journal of Transport Geography, 42, pp. 157–165.

Zazzi, M., Ventura, P., Caselli, B., & Carra, M. (2018). GIS-based monitoring and evaluation system as an urban planning tool to enhance the quality of pedestrian mobility in Parma. In M. Pezzagno, & M. Tira (Edited by), Town and Infrastructure Planning for Safety and Urban Quality (pp. 87–94). Leiden: CRC Press.

Sustainable, safe and resilient urban spaces

Pedestrians, Urban Spaces and Health – Tira, Pezzagno & Richiedei (eds)
© 2020 Taylor & Francis Group, London, ISBN 978-0-367-46171-3

Open source data and tools for disaster risk management: Definition of urban exposure index

R. De Lotto, C. Pietra & E.M. Venco
DICAr _ Department of Civil Engineering and Architecture, University of Pavia, Italy

ABSTRACT: Nowadays, natural hazards have huge social and economic impact in urban areas because urbanization and economic development increase people and physical and economic assets' concentration and therefore exposure in high-risk prone areas. Authors start from their previous researches related to urban exposure definition and urban function Exposure Index calculatron. Many scholars have outlined the importance of using geospatial technologies and up-to-date and accurate information from many different sources in all phases of Disaster Risk Managment using data collection also from the public, Volunteered Geographic Information. Authors identify the availability world-class open datasets related di DRM underlining differences and similarities among official data and VGI data, suggesting also how to implement maps and info necessary for the Exposure Analysis and DRM. At last, they provide a brief review of applied case study of VGI-open data-DRM in Lombardy Region.

1 INTRODUCTION

Natural hazards cannot be prevented but measures can be taken to mitigate their impacts and prevent them from becoming disasters. DRM represents the continuous process that aims to reduce the impact of natural hazards.

Nowadays, geospatial data and technologies are an integral part of DRM because both hazards and exposed society are changing in space and time (dynamic entities). In the pre-disaster phase, they contribute to support hazard and risk assessments through remote sensing data, ground surface models, census and statistical datasets using participatory web-GIS to spread and collect information among inhabitants. In the post-disaster phases, they play a major role in rapid damage assessment, also through crowdsourcing initiatives.

This potential often clashes with fragmentation, lack of harmonization, duplication of datasets, information and sources.

1.1 *Open data: A brief excursus*

Open data is data that can be freely used, reused and redistributed by anyone, subject only to the requirement to attribute and share alike (Open Knowledge Foundation, 2012). Public authorities are among the largest creators and collectors of data (geographic data, tourist information, statistical and business data, weather information, and so on), indispensable for policy development, public participation, and for decision-making process. The most important attributes of Open Data are (ePSI, 2014): availability and access; re-use and rdistribution; universal participation; interoperability.

The European Directive 2003/98/EC on the re-use of public sector information focuses on economic aspects of the re-use of information rather than on access to information by citizens. It encourages the Member States (at national, regional and local levels) to make as much information available for re-use as possible (Janssen, 2011).

The European Directive 2007/2/EC establishes INSPIRE (INfrastructure for SPatial InfoRmation in Europe): each European State must implement its own national environmental spatial data infrastructure, coordinating sub-national levels and providing geographic data, in order to guarantee more effective management; interoperability; sharing; abundance and usability; availability and access.

1.2 *Disaster risk management (DRM) and exposure index definition*

The Sendai Framework (2015) underlines the importance to reach an improvement of disaster risk understanding in all its dimensions (hazards, exposure and vulnerability of people and assets), and a strengthening of disaster risk governance and management. To reach these goals, mapping the territory, all risk's components, and disaster events is fundamental.

Since, in an urban area, human beings (and buildings) are the most exposed elements to risks, the assessment of the exposure of the different urban functions is one of the essential features of DRM. Mitigate risks acting on exposure means remove, reduce and control the quantity and quality of exposed items presented in territorial sensitive areas. It is essential to study the behavior, location and real distribution of population at an appropriate spatial and temporal scale. Therefore, it is develop an Exposure Analysis with the definition of the Exposure Index as function of each single urban function in relation to the urban block, the municipal territory or the spatial domain considered (trans/interscalarity) (De Lotto et al. 2016):

$EI = f$ (people, population age, hours of functions use, functions m^2)

2   AIM OF THE RESEARCH

The research aims to identify the availability of open data (i.e. maps of natural risks; urban fabric characteristic; population density-distribution-crowding); underline differences and similarities among official data and VGI data; suggest how to implement maps and info necessary for the Exposure Analysis and DRM; provide a brief review of applied case study of VGI-open data-DRM in Lombardy Region.

3   RESEARCH APPROACH

Considering the vastness of open and VGI data and the potential real-time and georeferenced territory understanding, the main research question is: there is reliable VGI data for urban exposure assessment and therefore risk? What kind of data is available? There are available datasets? In which contexts and to which scale of detail? The next paragraphs give a first answer regarding the type of data and how to treat it in relation to DRM.

4   APPLICATION

As natural hazards are location dependent, DRM can benefit from geographical representations. The latest developments in all areas of ICT and GIS-based software offer a huge potential for supporting the whole DRM lifecycle. Such systems are heterogeneous, and require a huge amount of spatial data (land models, road network vector-data and images, meteorological data flows, real-time measurements, land usage, infrastructures, real-time spatial distribution of population). Therefore, the most important characteristics for this data are: availability; standardization; security and reliability.

The EC defines 11 kinds of datasets useful in DRM (ePSI, 2014), listed below with some examples of world-class open datasets:

- Satellite imagery: COPERNICUS.
- Elevation and surface models: GDEM; OpenTopography.

- Meteorological data: WorldClim; European Climate Assessment and Dataset; Climate Analysis Indicators Tool.
- Transportation networks: Open Flights; World Port Index; Global Roads Open Access Data Set; Undersea Telecommunications Cables; Capitaine European Train Stations.
- Demographics and population density: Gridded Population of the World.
- Country and urban borderlines: GADM.
- Land use and buildings: Corine Land Cover Map; GLC-SHARE; Forest GIS; Mineral Resources Data System; Global Land Use Dataset; ESPON Urban Morphological Data.
- Utility networks: National Grid GIS data.
- Critical infrastructures.
- Vulnerable (sensitive) locations.
- Points of interest: POI Factory.

Moreover, as open dataset are also important those with maps of natural risks: USGS Earthquakes Database; Global Seismic Hazard Map; IBTrACS; MODIS Fire Detection Data; Natural Disaster Hotspot.

In relation to Exposure Analysis, the most important datasets for the definition of the Exposure Index are "Demographics and population density" and "Land use and buildings". From the first one, it is possible to deduce data related to the inhabitants present in the site considering census data as main source. Other data: density of inhabitants, age of population and type of occupation. From the lattes, the most important information are on urban (and territorial) functions (office, retail, facilities, and so on) and therefore the crowding indexes in the various functions (considering employees and users).

4.1 *Volunteered geographic information (VGI) and DRM*

It is very important using up-to-date and accurate information in all phases of DRM, and integrating information from many different sources including *in situ* sensors, aerial and satellite images, administrative, statistics and socioeconomic census data (Poser et al., 2010). New technologies have facilitated data collection from the public, introducing VGI in DRM that offers a great opportunity to enhance awareness because of the potentially large number of non-expert volunteers familiar with the specific area. They report the status changes observing important DRM parameters using mobile devices (smartphones, tablets, cameras) or *ad hoc* Web-GIS services defining geo-referenced and real-time information (Pollino et al., 2012).

DRM requires scientific, up-to-date data and scale appropriate hazard and exposure data: government usually provide this information. If the spatial data does not yet exist, some tools allow importing data from external source quickly and easily. The connection with Open Data and VGI datasets allows more detailed information to be collected, while being part of GIS environment allows easy spatial analysis (Pasi et al., 2015).

4.2 *Example of application in Lombardy region*

With particular reference to Lombardy Region, the use and dissemination of Opendata is possible thanks to the homonymous portal: the data reported are institutional and come directly from the State and government agencies (Ministries, ISTAT, INPS), regions (ARPA), municipalities. Lombardy Region has chosen to acquire a well-established service that offers a series of tools for using online data and not just a repository from which to download datasets.

The SIMULATOR-ADS Project (2017) represents a successful example of these data integration born to be applied on Lombardy territory. It was developed as a platform capable of supporting local authorities in identifying, forecasting, preventing, monitoring and managing risks of natural origin and anthropic origin. This approach can be considered both top-down and bottom-up. In fact, the local administrators have the possibility to communicate with citizenship by making available the prevention and emergency plans (official sources), and citizens can send useful observations to identify potential risk situations, allowing targeted interventions and consequent risk mitigation (unofficial sources). The platform also fully complies with regional, national and European legislation in the matter of territorial safety.

5 DISCUSSION

Considering official open data and VGI data, it is important to underline that at small to medium scales, there are no visual remarkable differences between the output layers produced: official and community data are comparable. Differences that are more significant can be detected at a larger scale: here there is a lack official data updating and individuals are not able to control elements in such wide scale territories.

An important advantage related to using VGI datasets (i.e. Open Street Maps) is the possibility of improving its data. In relation to DRM and Exposure Analysis, some of the elements that could be added to these datasets are: existence of underground floors; specific land use of each floors; approximate height at which the most vulnerable urban land uses are located, due to the buildings shape and structure; particular and/or valuable structural characteristics or changes regarding the structural part of the building; actual use or abandonment/closed areas for going-out-of-business/holidays/renovations of portions of buildings; any situations of renting/subletting for student or similar.

This allows to overcome the problem about medium-low resolution of available exposure data (information about buildings: from municipality to district units, and in very limited cases up to single block or building) that is not compatible with hazards maps and the level of detail necessary to DRM (spatial resolution non-homogeneity between hazard and exposure data) (Figueiredo et al., 2016).

Moreover, a big issue of VGI data is the validation: users can insert an untrusted report due to skill based errors, rule based errors and/or violations (false report). In order to validate the information received, it is possible to use *ex-ante* and *ex-post* techinques. The first one prevents the creation of low quality VGI reports; the latter removes or classifies reports after they have been created.

Finally, it is possible to affirme that the potentially infinite quantity of open VGI data requires specific tools for standardization, re-elaboration and analysis in DRM that are not so common at the moment.

6 CONCLUSIONS

As emerged, there is a huge amount of VGI data useful for Exposure Analysis and DRM. This data is compatible with official open data and therefore interoperability is guaranteed. Furthermore, the data reliability, once validation tests have been carried out, is very high as it is produced by locals who act as human sensors with a peculiar real-time perception of place and status changes. To improve its efficacy and to spread its potentiality in DRM, it is essential to produce realistic natural hazard impact scenarios using open and VGI data dialoguing together.

From the regulatory point of view is important define the rule structure in the different scale plans and also the level of detail useful for EI calculation and standardize data and methodology.

REFERENCES

De Lotto R. et al. (2016) Proposal to reduce natural risks: Analytic Network Process to evaluate efficiency of city planning strategies. GEOG-AN-MOD Proceedings, Part IV Editors: Gervasi O., Murgante B. et al. (Eds.) pp.650–664.

Directive 2003/98/EC of the European Parliament and of the Council of 17 November 2003 on the re-use of public sector information.

Directive 2007/2/EC of the European Parliament and of the Council of 14 March 2007 establishing an Infrastructure for Spatial Information in the European Community (INSPIRE).

Faravelli M. et al. (2017) Seismic damage scenarios at municipal level: application to the SIMULATOR project (Integrated module for risk management and prevention). Progettazione Sismica Vol. 8, N.3, 33–49.

Figueiredo R., Martina M. (2016) Using open building data in the development of exposure data sets for catastrophe risk modelling Nat. Hazards Earth Syst. Sci., 16, 417–429.

Goodchild M. F. (2007). Citizens as sensors: the world of Volunteered Geography. GeoJournal, 69 (4), 211–221.

Janssen K. (2011) The influence of the PSI directive on open government data: an overview of recent developments. Government Information Quarterly, 28(4),446–456.

Open Knowledge Foundation (2012) What is Open Data?.

Pasi R. et al. (2015) Open Community Data & Official Public Data in flood risk management: a comparison based on InaSAFE Geomatics Workbooks n. 12 – "FOSS4G Europe Como 2015".

Pollino M. et al. (2012) Collaborative Open Source Geospatial Tools and Maps Supporting the Response Planning to Disastrous Earthquake Events. Future Internet 2012, 4, 451–468.

Poser K., Dransch D. (2010). Volunteered geographic information for disaster management with application to rapid flood damage estimation. Geomatica, 64, 1, 89–98.

UNISDR (2015) Sendai Framework for Disaster Risk Reduction 2015 – 2030. Geneva.

Vescoukis V., Bratsas C. (2014) Open Data in Natural Hazards Management. ePSIplatform Topic Report No. 2014/01, January 2014.

Pedestrians, Urban Spaces and Health – Tira, Pezzagno & Richiedei (eds)
© 2020 Taylor & Francis Group, London, ISBN 978-0-367-46171-3

Strategies able to improve the level of safety and protection from seismic and hydro-geo-morphological risks

M.A. Bedini, F. Bronzini & G. Marinelli
Department Simau, Marche Polytechnic University, Ancona, Italy

ABSTRACT: The aim of this paper is to evaluate possible strategies to slow down the escape from fragile territories, favoring new settlement models based on interconnected urban systems, resilient and sustainable, end protected from risks.

It manifests itself with the acceleration of abandonment, in large territories, of hundreds of small widespread historical centers. To this widespread phenomenon the absence of strategies able to improve the level of safety and protection from hydro-geo-morphological and seismic risks is superimposed.

In some Italian regions, above all in Central Italy, there are new potentialities for a type of development, based on the rural dimension and a new production-settlement model: the so-called "productive landscape".

It is not only a question of substituting damaged assets and infrastructures, but also a question of reconstructing communities, to ensure equity, access to resources and equal opportunities for disadvantaged persons. Only in this way will it be possible to reduce the community's vulnerability to risks.

As already confirmed by positive experiences, different planning tools must be integrated, since the territory regeneration system involves a complex security, protection and maintenance system: SUM, Minimum Urban Structure to reduce urban seismic vulnerability, QSV, Strategic Valorisation Framework of Historical Centres, Civil Protection Plan, PAI, Hydrogeological Structure Plan, Seismic microzonation and thematic risk maps.

Keywords: resilient and sustainable models, fragile territories, hydro-geo-morphological and seismic risks

1 INTRODUCTION. THE PROBLEMS OF SUSTAINABLE AND RESILIENT URBAN SPACES

The fragility of the territories in Italy represents a problem of enormous dimensions, which seems extraneous to the priority choices of the political agenda.

It manifests itself with the acceleration of abandonment, in large territories, of hundreds of small widespread historical centers. Abandonment generated by a rapid reduction in the possibilities of work and livability, which has led to the departure of young residents and the natural disappearance of the older population.

To this widespread phenomenon the absence of strategies able to improve the level of safety and protection from seismic and hydro-geo-morphological risks is superimposed.

For example, on the occasion of the Rigopiano (Abruzzo) tragedy, it was once again observed, at a very high cost in terms of human lives, that buildings cannot be built on gullies at risk of avalanches, on water streams conveyed in large pipes, on precarious natural or artificial balconies, on lands at risk of landslides or hydrogeological disasters, or on areas that are unstable from a geotechnical point of view.

2 OBJECTIVES. USE OF INTEGRATED PLANNING TO SLOW DOWN THE ABANDONMENT OF FRAGILE TERRITORIES

In order to slow down the abandonment of fragile territories, this study proposes a new sustainable settlement model with greater protection from risks.

As already confirmed by field experimentation, different planning tools must be integrated, given that the territory regeneration system entails a more complex security, protection and maintenance system of the territory: SUM, Minimum Urban Structure to reduce urban seismic vulnerability, QSV, Strategic Valorisation Framework, Plan of Historical Centres, Civil Protection Plan, PAI, Hydrogeological Structure Plan, Seismic microzonation and thematic maps of risk.

Today we could say that the role of planning in influencing risk, vulnerability and/or exposure can be expressed by the following scheme (Figure 1) which shows the fundamental structure of risk.

Risk measurement can be obtained as a summation of the contribution of the Hazard, Exposure and Vulnerability.

Figure 2 shows the different elements of the Minimum Urban Structure of the Munipality of Bevagna (PG): structuring routes (urban crossings, links to the exterior and the main

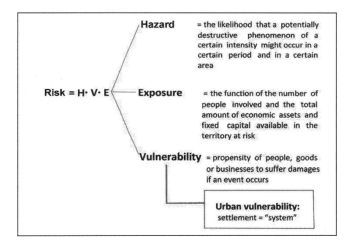

Figure 1. The fundamental structure of risk (Tira, 2017).

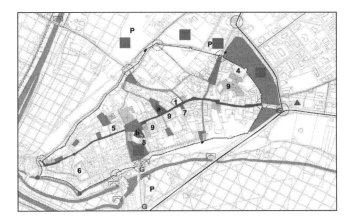

Figure 2. The different elements of the Minimum Urban Structure of the Munipality of Bevagna (PG) (Cappuccitti, 2017).

Figure 3.   An area in the Munipality of Campotosto (AQ), Abruzzo Region, earthquake 2016. The situation in March 2019.

functional and strategic centres), roads to junctions and gateways to the city), escape routes, secure areas, hospitals and civil protection centres and the fire brigade (Cappuccitti, 2017).

Figure 3 highlights the state of abandonment of an area in the Municipality of Campotosto (AQ) after 3 years from the 2016 earthquake.

In order to encourage residents to return (Alexander, 2013), it should be remembered that reconstruction is closely linked to the financial situation of citizens and businesses before the disaster, given that it is citizens with more money who are better equipped to face a disaster. And as researchers have observed, reconstruction may lead to a "boom-and-bust" economy, where the reconstruction process fuels a temporary economic growth, at the expense however of the long-term sustainability of the local economy.

That is why it is so important to choose which economy to support.

For example, the decision may be made to sustain the so-called "Productive Landscape" with a programme of strategic incentives (Bronzini, Bedini, 2015; Bedini, Bronzini, 2016).

The investment in rural settlements of landscape and environmental value may therefore be a driving force for development, for a new growth model based on the social and production reconversion of the countryside and the environmental and cultural revaluation of the net-work of widespread settlements, typical of the Central Italy.

In some Italian regions, above all in Central Italy, there are new potentialities for this type of development, based on the rural dimension and a new production-settlement model, such as that proposed by the rural policy of the CAP, the approach of Urban Agriculture (Torquati, Giacchè, Venanzi, 2015; Fleury, 2005) (Poulot, 2007) and the new Agropolitan forms (Donadieu, 2005; Droz, Forney, 2006), Agrourbanism (Fleury, Vidal, 2010), the Food Plans applied in Europe and the USA, or the urban horticulture projects (Daly, 2015) which appeared, in the late 90s, as a way to reorganise Irish agriculture.

In order to reconvert the agro-zootechnical industry, in line with food and production requirements, and social, tourist and cultural needs, agricultural planning should include the introduction of Wide-Area Plans.

This is confirmed by the successful experience of small groups of associated municipalities in Marche Region (in the hinterland of Pesaro) and other towns in Emilia Romagna, where it has been demonstrated that small Municipalities alone are not able to plan new and existing agricultural activities, that can be done only at a district level.

3 METHODOLOGICAL APPROACH: DIFFERENT LEVELS OF OPERATORS AND DIFFERENT STRATEGIES

An economic-territorial policy is therefore required with diversified operational tools: urban plans, farm park plans (Giacchè, 2014), agro-urban programmes (SDRIF) (agro-urban programmes to protect farming areas and encourage the active participation of farmers in territorial planning choices), *agriurbanism* projects (Maraccini *et al.*, 2013), Integrated Territorial Projects (PIT), (PSN, 2006), and tools to implement the structural funds of the European Union.

Consequently, in this moment of instability caused by the global and local crisis in Central Italy, and the difficult resurgence of lands devastated by the earthquake, the settlement filaments, the urbanised countryside and the hundreds of historic-rural buildings may represent an opportunity to relaunch local resources, expertises and lifestyles in areas of high environmental-landscape value with population widespread on the territory.

An effective strategy must be adjusted according to the endogenous potential of places, involving different levels of operators:

- local government authorities (responsible for protecting farm lands and functionality);
- groups of agricultural entrepreneurs (Milone, Ventura, 2009) (whose responsibility is to increase the quality of products and offer recreational, educational and social-environmental services);
- tourist or food and wine entrepreneurial groups (who also supply sports, social services, environmental recreation);
- cultural groups (in order to insert the historical rural and natural heritage in the economic network);
- artisanal businesses (with typical products);
- old and new local residents (with whom the preservation and functionality of places should be agreed).

This leads to a new planning approach of public spaces in the extensive territory, suggesting elements of recognisability, identification, the boundaries of the settlements and environmental qualification.

4 INNOVATIVE ASPECTS. OPERATIONAL STRATEGIES

This paper can only limit itself to proposing simple and clear procedural and planning suggestions, that could be used as a Road Map for public initiatives.

An earthquake breaks the existing fragile balance of the territory and makes it necessary to rethink the lifestyle model in these places, based on:

- a territorial urban risk protection system;
- a system of Smart land projects, that improve the services to the community and businesses in the territory;
- a system of new functional relationships between small urban centres in earthquake damaged areas and urban centres outside earthquake damaged areas (transport, cooperatives, itinerant services for production activities, etc.) and intangible relationships that strengthen the new ties between population resident in mountain, in hillside and in coastal areas.

The experiences in the past have produced successful in Emilia Romagna, Umbria, Marche, but also negative results in the historical centre of L'Aquila. In summary, there should be no Emergency Plan without Risk Prevention Plans. There should be no Emergency Plans with temporary settlements (which are often not really temporary) and the start of reconstructing buildings as a separate and distinct phase, if at the same time post-earthquake planning-management is not started.

5 RESULT OF THE RESEACH. THE POST-EARTHQUAKE PLANNING

In the fragile territories of Central Italy hit by earthquakes a new model of coexistence with seismic risks and environmental instabilities must be pursued, based on five key points:

a. the introduction of all the most advanced technologies in order to make these territories less isolated, less abandoned, and no longer heading towards a terminal phase of decline: from broadband networks to the remote management of public and private services, energy control, the reorganisation of the waste system and slow mobility for the re-appropriation of the values of the green heart of Italy. An advanced monitoring system (Gasparini *et al.*, 2007) that alerts residents and the emergency services of the potential cascading effects of earthquakes, landslides and flooding.
b. the localisation of territorial "hubs", i.e. "functional equipped pivots", at the service of extensive settlements and the conservation or re-localisation of primary services (schools, health care facilities, public offices, etc.);
c. the identification of new business areas, and new protected areas to shelter animals;
d. the recovery, through social *housing* and self-reconstruction practices, of the small and very small historical centres that are scattered throughout the area, linked by a dense web of farm paths;
e. the identification, on detailed maps, of buildings, public spaces and protected routes, that, in case of natural disasters, ensure internal and external access and the continued efficiency of network services.
f. the stable integration of "new social components", through strategic public programmes, which are conducive to creating a new economic development model. A model concentrated on the resources available, resources that are under-used or badly used, cultural tourism, local production of food and wine, modern agro-zootechnics, creative artistic activities, alternative ways of recovering buildings in new economic circuits (Campos Venuti, 1980).

6 CONCLUSIONS. THE RESURGENCE OF FRAGILE TERRITORIES

From what has been explained up to now, the following strategic choices can be considered consolidated acquisitions by the scientific community to live with earthquakes in the coming decades: a clear and decided

"Yes" to the main great objective: to bring back to the devastated places the population and activities present before the earthquake.

"Yes" to the Masterplan, intended as a programmatic tool in order to demonstrate the complexity of the urban planning regeneration strategy.

"Yes" to simplification/integration of several interrelated Plans (SUM, Civil Protection Plans, Hydrogeological Defence Plans, Seismic Microzonation Plans).

"Yes" to planning multi-scale Minimal Urban Structures, at an urban and Wide-Area level, with forms of governance that include several associated Municipalities.

"Yes" to Area Plans at the neighborhood and urban level and Wide-Area Plans, in order to plan the reorganisation of a large urban *widespread* system, where the focus is on the economic, social, cultural and productive resurgence of the areas affected.

"Yes" to the provision of advanced network technologies to allow coexistence with the earthquake in areas spread over large territory, ensuring the maximum protection, assistance, evacuation in case of recurrence of disasters.

"No" to focusing exclusively on reconstruction in post-earthquake interventions.

"No" to build new settlements, declared improperly temporary; in non-building areas, intended for environmental protection, by way of derogation from urban planning tools.

"No" to the uncontrolled delegating for the reconstruction to the mayors of small Municipalities, with little resources, experience and skills and many local conditionings, even illegitimate.

"No" to encourage the purchase, by residents of the affected areas., of new housing away from the earthquake zones.

"No" to abandoning home owners to a multitude of engineers, interested only in taking professional assignements.

REFERENCES

Alexander D., (2013). Planning for Post-Disaster Reconstruction. Available at: http://www.grif.umon treal.ca/pages/papers2004/Paper%20%20Alexander%20D.pdf.

Bedini M.A., Bronzini F., (2016). The New Territories of Urban Planning. The Issue of the Fringe Areas and Settlements. Land Use Policy. The International Journal Covering All Aspects of Land Use 57. 130–138.

Bronzini F., Bedini M.A., (2015). The embrace of the city countryside. Archivio di Studi Urbani e Regionali 112. Franco Angeli, Milan. 60–76.

Cappuccitti A., (2017). Mitigare le vulnerabilità territoriali, integrare gli strumenti urbanistici [Mitigating territorial vulnerabilities, integrating urban planning tools]. Lesson at Master "City and Territory", Ancona, June. Programma preliminare del Quadro Strategico di Valorizzazione del Centro storico, Comune di Bevagna (PG), Capogruppo Prof. G. Imbesi.

Campos Venuti G., (1980). An indifferent economy in the territory is not possible. Rinascita, 48.

Daly S., (2015). Producing healthy outcomes in a rural productive space. Journal of Rural Studies 40. 21–29.

Donadieu P., (2005). From utopia to the reality of urban campaigns. Urbanistica 128.

Droz Y., Forney J., (2006). Quelles perspectives pour les "Exclus du terroir"? Le cas des exploitations agricoles du Canton de Neuchâtel [What perspectives for the "Excluded from the soil"? The case of farms in the Canton of Neuchâtel]. In: Conférence/Débat à Agropolis Museum, Montpellier, 29 November. Available at: http://www.museum.agropolis.fr/pages/savoirs/exclusterroir/forney_droz_2006.pdf.

Fleury A., (2005). The construction of agri-urban territories in Ile-de-France. Urbanistica 128. 20–24.

Fleury A., Vidal R., (2010). L'autosuffisance agricole des villes, una vaine utopie? [The agricultural self-sufficiency of cities, a vain utopia?]. In: La vie des idées. Available at: http://www.laviedesidees.fr/IMG/pdf/20100604_villesdurables_vidal_fleury.pdf.

Gasparini P., Manfredi G., Zschau J. (Eds.), (2007). Earthquake early warning systems. Springer, Berlin.

Giacchè G., (2014). L'expérience des parcs agricoles en Italie et en Espagna: vers un outil de projet et de gouvernance de l'agriculture en zone périurbaine [The experience of agricultural parks in Italy and Spain: towards a project and governance tool for agriculture in peri-urban areas]. In Lardon S., Loudiyi S. (Eds.), Agriculture urbaine et alimentation: entre politiques publiques et initiatives locales. Revue Géocarrefour 89. 1–2.

Maraccini E., Lardon S., Loudiyi S, Giacchè G., Bonari E., (2013). Durabilité de l'agriculture dans les territoires périurbains méditerranéens: enjeux et projets agri-urbains dans la région de Pise (Toscane, Italie) [Sustainability of agriculture in the Mediterranean peri-urban territories: challenges and agri-urban projects in the region of Pisa (Tuscany, Italy)]. Cahiers Agricultures 22, 6. 1–9.

Milone P., Ventura F., (2009). The farmers of the Third Millennium. Behaviours, Expectations, Proposals. Publisher AMP, Perugia.

Poulot, M., (2007). De la clôture patrimoniale des territoires périurbains dans l'ouest francilien [Heritage closing peri-urban areas in the western Paris region]. Socio-anthropologie 19.

PSN Strategic National Plan for Rural Development (art. 11 Ec Reg. 1698/2005), (2006). Ministry of Agricultural and Forestry Policies, Rome.

Tira M., (2017). Pianificazione urbanistica e mitigazione del rischio [Urban planning and risk mitigation]. Lesson held at the Master City and Territory. Innovative Strategies and Tools for Risk Protection of Territories in Crisis, Camerino, July.

Torquati B., Giacchè G., Venanzi S., (2015). Economic Analysis of the Traditional Cultural Vineyard Landscapes in Italy. Journal of Rural Studies 39. 122–132.

Pedestrians, Urban Spaces and Health – Tira, Pezzagno & Richiedei (eds)
© 2020 Taylor & Francis Group, London, ISBN 978-0-367-46171-3

Flood vulnerability functions for people and vehicles in urban areas

M. Pilotti & L. Milanesi

Department of Civil, Environmental, Architectural Engineering and Mathematics, University of Brescia, Italy

ABSTRACT: Flood risk assessment in urban areas is a key aspect for land use planning and emergency management. Most of the flood-related fatalities are related to unaware behaviours and erroneous risk perception of people entering flooded ways or driving vehicles. Accordingly, rational vulnerability criteria for people and vehicles under the action of flood flows are a prerequisite of risk assessment.

This paper presents the operational formulations of two state-of-the-art physically based vulnerability models for people and vehicles impacted by a flow. The proposed vulnerability models are applied separately to the numerical results of a flood propagation model in urban area. The resulting risk maps allow to identify the unsafe areas for people and vehicles and may guide the setup of proper evacuation strategies and emergency plans.

1 INTRODUCTION

Floods are the most relevant natural hazard worldwide in terms of victims and costs and their impacts are expected to grow following the socio-economic development and the intensification of the hydrological cycle triggered by climate change. Flood risk assessment requires accurate hazard models, exposure metrics and vulnerability functions for the most relevant targets. Although advanced numerical methods are available for the first task, flood vulnerability models describing the interaction between the hazard metrics and the exposed elements are still far from being applied with homogeneous and widely accepted criteria. Focusing on floods in urban areas, people (e.g. Milanesi et al., 2014, 2015 and 2016), structures (e.g. Milanesi et al., 2018), and vehicles (e.g., Milanesi & Pilotti, 2019) can be considered among the most exposed elements.

Reviews on flood fatalities during the last century in the U.S. (e.g., Ashley and Ashley, 2008), Europe (e.g., Jonkman and Vrijing, 2008) and Australia (e.g., Haynes et al., 2016) showed that their majority is vehicles-related and an additional fraction is due to people intentionally entering currents and swept away by the flow. Accordingly, both the direct vulnerability of people and of vehicles (that can cause damages to drivers and pedestrians), are worth of being considered when the risk assessment is primarily focused on human safety. In this case, the choice of physically based vulnerability criteria, whose meaning is clearly understandable by stakeholders and policy makers, is crucial to increase the effectiveness of the protection measures.

Two simple and weakly parameteric literature models assessing the stability of a person (Milanesi et al., 2015) and of a vehicle (Milanesi & Pilotti, 2019) are briefly discussed. Both models have a sound physical basis since they include the geometrical conceptualization of the impacted objects (i.e. human body and vehicle) and of the involved forces. These models can be used for evaluating the average stability condition of the population of different geographic areas and of a large variety of real vehicles. This approach based on objective vulnerabiliy models provides a reliable support for emergency planning since it allows the identification of the most dangerous roads for walking people and for vehicles and the setup of proper mitigation strategies such as safe evacuation routes and passing interdictions. Both models currently provide the best fit to the available literature experimental data set. Finally, simplified formulations of the two vulnerability models are provided in order to foster their application for operational purposes.

As an example of application, the vulnerability models are applied to the 2D flood model of the conceptualized urban district test case proposed by Testa et al. (2007) assuming a synthetic discharge hydrograph. The vulnerabiltiy functions are applied both to the maps of maximum depth and velocity and to the instantaneous depth and velocity fields in order to discuss the influence of the non-contemporaneity of depth and velocity maxima on the resulting risk maps.

2 METHODS

2.1 *People vulnerability to floods*

A detailed explanation of this model is provided in Milanesi et al. (2015). The human body is represented in vertical position on an inclined slope and impacted frontally by a steady uniform flow. The body weight is decomposed in direction normal and parallel to the slope. Considering the pressure distribution of a parallel flow, the buoyancy force is normal to the bed. Since the body stands in vertical position on the slope, it is not normal to the flow field. The fluid dynamic force is decomposed in direction parallel (drag) and normal (lift) to the slope. Friction between the soles and the bed is the product of the coefficient μ (0.46, after calibration) and the effective weight that is the algebraic sum of the forces normal to the slope (normal component of weight, buoyancy and lift). Three different critical conditions are considered: slipping instability occurs if the sum of the drag force and of the weight parallel component overcomes friction; toppling instability occurs if the moment, calculated with respect to the heel, of the normal component of the weight is exceeded by the destabilizing moments due to lift, drag, buoyancy and the parallel component of the weight; drowning is considered by imposing a maximum admissible water depth as a function of the height of the neck. The depth h safety limit, as a function of the flow velocity U, is given by the minimum among slipping, toppling and drowning depth. The geometrical parameters describing the human body were retrieved from medical studies. Two set of parameters representative of European children and adults were used to compute the thresholds in Figure 1a. For the case of water on a horizontal slope, two easy to use operational formulations of these thresholds can be obtained through interpolation as:

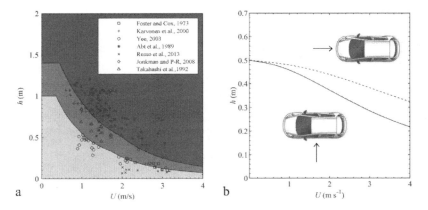

Figure 1. (a) Stability thresholds for adults (upper line) and children (lower line) for the case of water flow on horizontal terrain; dots represent the literature experimental data. High, medium, and low vulnerability areas are identified, respectively, by red, orange, and yellow shading. (b) Stability thresholds for the reference vehicle RV2 (Milanesi & Pilotti, 2019) impacted by flow in parallel (dashed line) and perpendicular (solid line) direction.

$$h_{adults} = \min\left(1.4; \frac{2.809}{U + 1.102} - 0.416\right) \tag{1}$$

$$h_{children} = \min\left(1; \frac{1.565}{U + 0.884} - 0.275\right) \tag{2}$$

2.2 Vehicles vulnerability to floods

The vehicle is conceptualized as a rectangular prism on an inclined slope and impacted by a steady uniform flow. The prism is detached from the ground of a height G_C (Ground Clearance). Weight and buoyancy act as in the case of the people vulnerability model and can be computed on the basis of the geometry of the vehicle. Drag and lift forces are computed with respect to the transversal and planform area of the vehicle, respectively. Friction between the tires and the bed is the product of the coefficient μ that in general can be variable as a function of the relative orientation of the vehicle. For application purposes, a single value $\mu=0.6$ will be considered in the following. The only failure mechanism to be considered in the case of a stationary vehicles is slipping instability that occurs if the sum of the drag force and of the weight parallel component overcomes friction. The computation of the lift and drag coefficients was based on a regressive analysis of the available experimental data that allowed to obtain a set of parameters valid to represent the average stability condition of a wide variety of circulating vehicles. The geometric parameters of the vehicle are representative of the average features of the circulating European vehicles. The undisturbed U stability limit, as a function of the flow velocity depth h and of the local slope ϑ, is given, for parallel and perpendicuar relative orientation of the flow respectively, as:

$$U \leq \left[\frac{(604 - 685\tan\vartheta - 1218h)\cos\vartheta}{38h + 1}\right]^{1/2} \tag{3}$$

$$U \leq \left[\frac{(390 - 443\tan\vartheta - 788h)\cos\vartheta}{68h - 1}\right]^{1/2} \tag{4}$$

Equations 3 and 4 are defined for $G_C < h < h_b$, where h_b (here $G_C = 0.158$ m $h_b \approx 0.495$ m) is the depth at which the vehicle starts floating, and represent the numerical formula of the stabilty conditions to sliding for the case of a water current as detailed in Milanesi & Pilotti (2019). The stability threshold for stationary vehicles impacted by a flow in perpendicular orientation is more cautionary than the case of parallel direction (Figure 1b). Accordingly, for safety reasons the first one will be considered in the following analyses.

3   RESULT OF THE RESEACH

The propagation of a synthethic flood wave over the domain of the simplified urban district 1:100 scaled physical model in the Toce valley proposed by Testa et al. (2007) allowed to discuss the effect of different vulnerability criteria on the resulting risk maps. The fully 2D simulation was performed with Hec-Ras 2D implementing the full momentum form of the Saint-Venant equations. The building were treated as rigid structures and the roughness $n=0.0162$ s m$^{-1/3}$ was assumed (Testa et al., 2007). A mesh with mean size 0.05 m was adopted and locally refined with cells of size up to 0.01 m near the buildings in order to best represent their interaction with the flow. Given that the simulations were performed in the model scale, the resulting depth and velocity maps were scaled to the prototype dimension for the application of the vulnerability models assuming the scale ratios dictated by the Froude similitude.

As customary in flood risk assessment, vulnerability criteria are applied to the depth and velocity maximum values that, in general, do not occur at the same time. As an example, the

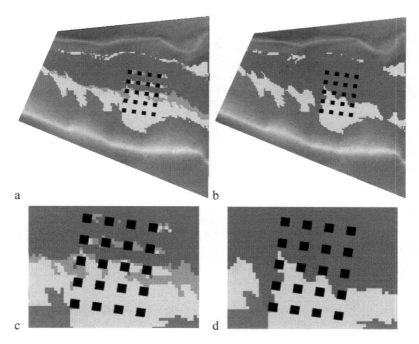

a   b

c   d

Figure 2. Envelope risk maps for people (a) and vehicles (b) computed using instantaneous flow depth and velocity values. Detail of the risk maps for people (c) and vehicles (d) inside the urban area computed with non-contemporaneous flow depth and velocity maxima.

1D case of a reflected wave impacting a rigid wall is emblematic. Accordingly, it would be more appropriate to apply vulnerability models at each computational timestep to build an envelope map of the maxima of the resulting risk values. However, the use of commercial codes often hinders this approach. Figure 2 shows the risk maps for vehicles and people computed with both these approaches. Since the vehicles vulnerability function is more cautionary than the people stability thresholds, the related risk map (Figure 2b) shows a larger high-risk area than the people risk map (Figure 2a). In both cases, the risk maps computed with the depth and velocity maxima (Figures 2c and 2d) are more cautionary than the envelope scenarios described above. In this case, however, the differences are negligible.

4 CONCLUSIONS

Flood risk maps based on rational vulnerability models are powerful tools for the identification of effective risk mitigation measures. Focusing on the definition of emergency plans, vulnerability thresholds of the stability of people and vehicles impacted by a flow allow to identify proper strategies to minimize the threat to the population. The example shown in this paper demonstrates the effectiveness of these models in the identification of the most dangerous areas for pedestrians and vehicles. Finally, the aspect of the computation of risk maps based on contemporaneous values fo flow depth and velocity was assessed.

REFERENCES

Ashley, S. T., & Ashley, W. S. (2008). Flood fatalities in the United States. Journal of Applied Meteorology and Climatology, 47(3), 805–818.
Jonkman, S. N., & Vrijling, J. K. (2008). Loss of life due to floods. Journal of Flood Risk Management, 1(1), 43–56.

Haynes, K., Coates, L., Dimer de Oliveira, F., Gissing, A., Bird, D., van den Honert, R., Radford, D., D'Arcy, R., & Smith, C. (2016). An analysis of human fatalities from floods in Australia 1900-2015. Report for the Bushfire and Natural Hazards CRC, Australia.

Milanesi L., & Pilotti, M. (2019). A conceptual model of vehicles stability in flood flows. J. Hydraul. Res., in press.

Milanesi, L., Pilotti, M., & Bacchi, B. (2016). Using web-based observations to identify thresholds of a person's stability in a flow. Water Resources Research, 52(10), 7793–7805.

Milanesi L., Pilotti M., Belleri A., Marini A., and Fuchs S. (2018). Vulnerability to flash floods: a simplified structural model for masonry buildings. Water Resour. Res, 54(10), 7177–7197.

Milanesi, L., Pilotti, M., and Ranzi, R. (2015). A conceptual model of people's vulnerability to floods. Water Resour. Res., 51(1), 182–197.

Milanesi, L., Pilotti, M., Ranzi, R., & Valerio, G. (2014). Methodologies for hydraulic hazard mapping in alluvial fan areas. Proceedings of the International Association of Hydrological Sciences, 364, 267–272.

Testa, G., Zuccalà, D., Alcrudo, F., Mulet, J., and Soares Frazao, S. (2007). Flash flood flow experiment in a simplified urban district. J. Hydraul. Res., 45(extra issue), 37–44.

Pedestrians, Urban Spaces and Health – Tira, Pezzagno & Richiedei (eds)
© *2020 Taylor & Francis Group, London, ISBN 978-0-367-46171-3*

The reduction of the population's residential exposure to radon risk in the municipal urban plan

R. Gerundo, M. Grimaldi & A. Marra
Department of Civil Engineering, University of Salerno, Fisciano, Italy

ABSTRACT: The exposure to radon in homes is one of the most important causes of lung cancer deaths worldwide. In view of the significant health effects, radon should be considered among the natural risks that urban and territorial planning has to manage, which suggests the possibility of recognizing different lines of action, corresponding to the various levels of planning. The proposed methodology, applied to the municipality of Eboli (SA), aims to reduce the population's residential exposure to radon risk in the municipal urban plan. To this purpose the plan addresses, the mitigation of risk levels in existing settlements on the one hand, and the prevention of popolation's exposure in new buildings of future settlements on the other, by requiring differentiated protection levels according to the risk and hazard levels respectively. Finally, specific guidelines to be attached to the plan indicate how to implement protection levels required.

1 INTRODUCTION

The exposure to radon in homes[1] is one of the most important causes of lung cancer deaths worldwide. The World Health Organization (Who) recommends including in the national building regulations, and/or building codes, prescriptions regarding the prevention and protection of the population from the radon risk (Who, 2009) and to identify areas where concentrations higher than the reference level[2] established at national level are expected, known from the literature as *radon prone areas*. In some countries, such as England, depending on the values found in these maps, different levels of protection are provided in the national building regulation, which correspond to simple interventions or more advanced interventions, as proposed by the British Research Establishment (Bre). In such areas, the Who recommends that more stringent measures should be provided. These recommendations have been fully incorporated into the recent European Directive of 2013, which obliges Member States for the first time to provide for protective measures against radon risk in homes. (EU, 2013). In Italy, late in transposing the European directive, specific legislation does not exist for homes, but only for workplaces and schools, which is Legislative Decree 241/00. Despite the regulatory gap, some regions have produced maps following the approaches used in Europe, on a regional scale.

An analysis of municipal building regulations conducted by the authors shows a general lack of attention to radon risk in urban planning, but when it is present, the reference to radon prone areas on a regional scale determines a uniform protectionist approach for buildings of the same municipality, due to the resolution of such maps, often on a municipal basis. Within this framework, it is advisable to define different levels of knowledge of radon risk, from the regional scale to the municipal scale, to support respectively regional and urban planning (Guida et alii, 2008).

---

1. Radon is a radioactive gas produced in the subsoil, which exhaling enters buildings where it reaches high concentrations, while outdoor it is dispersed by air currents.
2. According to the WHO, the optimum reference level is 100 Bq/m^3 and it cannot exceed 300 Bq/m^3 in any case.

2 OBJECTIVE OF THE PAPER

The paper shows a methodology to reduce the exposure of the population to radon in homes, both future and existing, through the municipal urban plan, then on an urban scale.The proposed methodology was applied to the municipality of Eboli, in the province of Salerno (Italy), as part of the technical and scientific support offered by the civil engineering department of the University of Salerno for the the municipal urban plan, currently in the drafting phase.

3 METHODOLOGICAL APPROACH

The methodological approach pursued provides for a differentiated protection of the population according to the different levels of hazard and risk in the municipal territory, summarizing the approach proposed by Bre, with the difference that it is applied on an urban scale, within the drafting of the municipal urban plan and in the framework of the general theory of territorial risk.

4 DESCRIPTION

The proposed methodology envisages two lines of action:

- Mitigation, by reducing population's exposure to radon in dwellings of existing settlements;
- Prevention, by reducing population's exposure to radon in new dwellings of future settlements.

For this purpose it is necessary the previous construction of the hazard and radon risk maps at the urban scale, after having clarified the fundamental equation of risk (R), that is expressed as the product of hazard (H), vulnerability (V) and exposure (E).

Radon hazard in dwellings is expressed as a function of both soil hazard, defined as the concentration of radon potentially exhaled from the soil, and the hazard from building, particularly form the construction characteristics of the building depending on which the passage through the building of the radon from the ground is more or less hindered. The vulnerability, thought in terms of resistance, is set equal to unity, a condition equivalent to that of maximum damage, since the exposed good, represented by the population, is incapable of limiting its exposure to an odorless and colorless gas such as radon.

In the absence of international mapping standards, an hazard and risk mapping methodology was proposed by the authors in a previous contribution, to which reference is made for further information (Gerundo et alii, 2016). Given this premise, with regard to mitigation, the method provides for the definition of three levels of protection, differentiated according to the intensity of the risk, derived from the risk map on an urban scale, classified according to five classes: low risk (R1), risk low (R2), medium risk (R3), high risk (R4), very high risk (R5). Specifically, for classes R1 and R2, no specific protection is provided; for class R3, basic protection is provided; for classes R4 and R5, advanced protection is provided. The protection levels correspond to specific intervention techniques, selected on the basis of the experience gained in European countries and in Italy[3]. The mitigation actions based on the risk levels identified in the final risk map, which constitutes a supplementary graphic drawing of the plan, are formalized in a specific section in the municipal building regulation. Guidelines which illustrate how to apply the mitigation techniques are also attached to building

---

3. These techniques are classified into passive and active techniques, depending on whether or not they require the external energy supply to the system. Examples of passive techniques are: joint sealing, natural ventilation. Examples of active techniques are: forced ventilation, depressurization, air conditioning with heat recovery (Stuk, 2012).

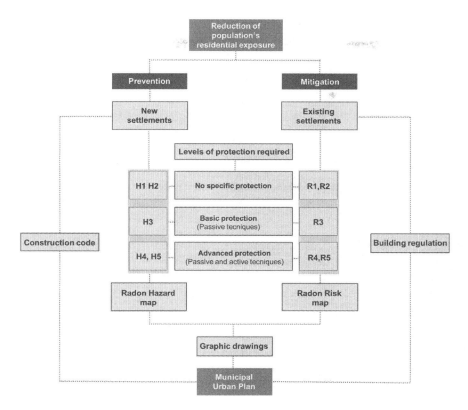

Figure 1. Methodological framework.

regulation, supporting the designers. With regard to prevention, the method provides for the definition of three differentiated levels of protection, depending on the intensity of the hazard from the ground, derived from the urban scale hazard map, classified according to five classes, from low hazard (P1) to hazard very high (P5). The map of the hazard from the soil, obtained with the methodology mentioned above, is a support tool for the definition of prevention actions already in the planning of future building developments. In particular, it allows to establish the severity of the necessary constructive measures, to be specified in the technical implementing regulations that govern the areas of transformation. Specifically, with reference to the different levels of hazard, the same three levels of protection for the risk mitigation in existing buildings are provided, with the difference that active prevention techniques may differ from those for existing buildings, while passive intervention techniques are essentially the same (Figure 1).

5 RESULTS AND DISCUSSION

The Municipality of Eboli presents a housing-building heritage largely concentrated in the main inhabited center, developed starting from the ancient center, born on the slopes of the hilly-mountainous part of its territory. In the flat area towards the Tyrrhenian coast there are three small inhabited areas and a considerable settlement dispersion. The drafting of the municipal urban plan is currently in progress, but the preliminary plan has already been approved.

With reference to mitigation actions, the application of the proposed methodology, foresees to prescribe in the building regulations of the drafting plan, an advanced protection in the buildings of some neighborhouds of the historic center and some newly formed residential

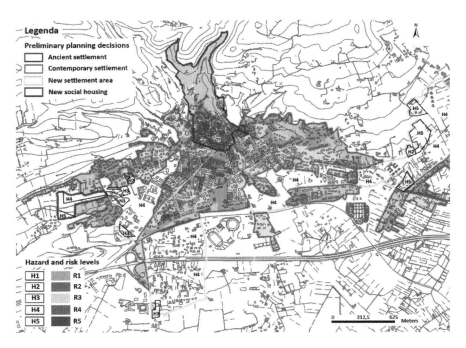

Figure 2.   Hazard and risk levels in new and existing settlements.

neighborhoods belonging to the main inhabited area, falling in areas of high and very high risk according to the radon risk map (Figure 2). However, much of the municipal building stock falls into areas of medium risk, for which basic protection is required. For the low-risk areas, where the three inhabited areas fall, no specific protection is prescribed, but simple good construction tricks that allow the isolation of the first treading are recommended. In this way it was possible to optimize resources, differentiating the levels of protection required within the municipal territory and thus giving priority to areas where a greater risk is expected. With reference to the new buildings, the application of the method made it possible to simulate the hazard level expected in the areas where transformation is allowed. On the basis of the hazard level from the soil hazard map, strict constructive measures will be prescribed in the technical implementation standards, since most of the areas that can be transformed for new settlements are affected by high and very high hazard levels. In this way a higher level of protection is assured to the population that will inhabit the new buildings, rather than providing simple constructive measures in all new dwellings, which may not be sufficient where the hazard level is significant.

6   CONCLUSIONS

In order to implement the radon risk among the canonically considered risks in urban and regional planning, the proposed methodology focuses on the municipal urban plan. The plan addresses, the mitigation of risk levels in existing settlements on the one hand, and the prevention of popolation's exposure in new buildings of future settlements on the other, by requiring differentiated protection levels according to the risk and hazard levels respectively. Finally, specific guidelines to be attached to the plan indicate how to implement protection levels required.

Future developments, with reference to the existing settlements, can be the definition of reward mechanisms that allow to reduce the levels of exposure through the recognition of building credits, relative to the application of mitigation measures, also in combination with interventions for protection from the seismic risk and energy efficiency. This should also be one of the main objectives of urban regeneration interventions to ensure healthy and resilient cities, capable of facing the threats to which they are increasingly exposed.

REFERENCES

EU-European Union, (2013). 2013/59/Euratom.
Gerundo R., Grimaldi M., Marra A., (2016). A methodology hazard-based for the mitigation of the radon risk in the urban planning. UPLanD – Journal of Urban Planning, Landscape & environmental Design. 1(1). 27–38.
Guida D., Guida M., Cuomo A., Guadagnuolo D., Siervo V., (2008) Radon-prone Areas Assessment in Campania Region. Applications of a hierarchical and multiscale approach to the environmental planning. Journal of technical and environmental geology.2/08. 38–62.
Stuk, (2012). RADPAR WP6, D13/1: Assessment of current techniques used for reduction of indoor radon concentration in existing and new houses.
WHO-World Health Organisation, (2009). Who Handbook on Indoor Radon: a public health perspective. Zeeb H. & Shannoun F. (Eds.).

Pedestrians, Urban Spaces and Health – Tira, Pezzagno & Richiedei (eds)
© 2020 Taylor & Francis Group, London, ISBN 978-0-367-46171-3

Assessing the economic and health impact of soft mobility. The Lombardy region case study

E. Turrini, C. Carnevale, E. De Angelis & M. Volta
DIMI, B+ Labnet – University of Brescia, Italy

ABSTRACT: Despite the political and scientific concern arisen in recent years, Particulate Matter (PM), is still a major problem especially in urban areas. Here, end-of-pipe measures are usually not enough to reduce atmospheric pollutant concentrations to acceptable levels, so, behavioral measures are a viable option to further abate PM precursor emissions. Soft mobility (SM), i.e. cycling or walking, is one of the most studied behavioral policies, due to its ability to generate multiple positive environmental and health impacts. This work presents the assessment of different soft mobility scenarios through the MAQ modelling system. The evaluation integrates fuel savings, greenhouse gases emission reduction, direct and indirect health impacts of SM measures and their implementation cost. The methodology is tested on a case-study for Lombardy Region in Northern Italy, a densely urbanized area with extremely high particulate matter concentrations.

1 INTRODUCTION/STATE OF THE ART

Particulate Matter (PM) and other pollutant characterized by a secondary fraction, i.e. O_3 and NO_2, have been demonstrated to cause adverse health effects on the population (WHO, 2013). To reduce these losses by preventing diseases and deaths, WHO Member States recently adopted a road map (WHO, 2018) that, in Europe, supplements the European Air Quality Directive (2008/50/EC), providing a framework to control outdoor PM concentrations. At urban scale, air quality plans could benefit from behavioral measures, encouraging, for example, Soft Mobility measures.

The European legislation already provides high levels of application for end-of-pipe measures for the coming years, making a further increase of their application, possibly not as efficient as the introduction of new strategies. These new strategies could be energy saving and behavioral measures, affecting directly fuel consumption, resulting in multiple positive effects on pollutant emissions, greenhouse gases emissions, monetary savings, direct and indirect positive health impacts. Among behavioral strategies, soft mobility measures (including walking and cycling) are a well-known and studied subject (Mueller et al., 2017) often addressed also at political level (WHO, 2017). Tainio et al., 2016 dealt with positive and negative health impacts of SM strategies on individuals due respectively to increased physical activity and increased PM exposure. WHO, 2014 developed a tool (HEAT) to economically assess the direct impacts of walking and cycling. Guariso & Malvestiti, 2017 evaluated the costs and benefits of various scenarios characterized by different amounts of people cycling. Arem, Moore, & Patel, 2015 and Mueller et al., 2017 quantified the dose-response function to link leisure time physical activity and mortality. Woodcock et al., 2018 developed a tool to assess the impacts of an increase in the cycling population on travel patterns, GHG emissions and health. The aim of this work is to present a methodology integrating the evaluation of direct health impacts due to SM application in the MAQ system (Turrini et al., 2018, Claudio Carnevale et al., 2018), an Integrated Assessment modelling tool design to evaluate costs and benefit of air quality policies and, in particular, in this paper, soft mobility. A test case on Lombardy region in Northern Italy, a densely populated regional domain with high PM concentrations and high road traffic levels, is presented.

2 METHODOLOGICAL APPROACH

MAQ is a Decision Support System that implements the DPSIR framework (EEA, 1999, Guariso et al. 2016) in a set of models and databases. It allows to identify the measures (Response) that directly abate precursors emissions (Pressures) or reduce the human activities (Drivers) causing a reduction of the air pollution level (State) which Impacts on human health, climate change, ecosystem and economy. The MAQ model considers two different types of emission abatement measures: (1) end-of-pipe measures, that abate the pollutant emission before they are released in atmosphere, with no modification of the energy consumption (e.g. filters applied to cars) and (2) energy efficiency measures, that modify the activity (drivers) causing pollution, changing the energy consumption. Such measures include both technological (e.g. substitution of fireplaces with gas boilers for domestic heating) and behavioral measures (e.g. soft mobility).

The decision variables of the problem implemented in MAQ are the application rates of these measures. The model allows two different approaches: scenario analysis and multi-objective optimization . In the scenario analysis the application rates of the measures are defined by the decision-maker and the impacts are then estimated by the system. In the optimization approach, one or more Air Quality Indexes and the policy implementation costs are minimized.

The MAQ system assesses direct and indirect health (and monetary) impacts of a set of measures. The indirect impacts are related to (i) the health impacts and external costs due to PM10 population exposure (ExternE, European Commission, 2005) and (ii) the increased breathing for commuters that choose to move from private cars to walking and biking (Tainio et al., 2016). The direct impacts are related to the benefit of the increase of physical activity of commuters that choose soft mobility (WHO, 2014). In this work MAQ model is used in scenario mode to assess the costs and the impacts on air quality of soft mobility scenarios.

3 CASE STUDY SETUP

The domain considered in this work is Lombardy Region. The Po valley area is a densely populated area with many industrial and logistic activities. The air quality levels in the region are a combination of population density, presence of industrial and logistic activities and low wind speeds.

The regional area has been divided into 1166 regular cells with a 6x6 km^2 area. Three different soft mobility measures have been included:

- walking for 20 min/day;
- commuting by bus, which implies 20 min/day of walking from and to bus stops;
- cycling for 40 min/day.

These measures are applied only to the commuting population.

The economic impact of these measures has been estimated adopting, two different approaches: in the first one it has been assumed as the institutional communication cost to publicize the measures, in the second approach a methodology assigning the monetary value of 14.1 €/hr to travel time (Asensio & Matas, 2008) has been applied .

4 RESULT AND DISCUSSION

Four scenarios have been compared:

- the Base-case, in which a cost-efficient set of end-of-pipe measures and energy measures is applied on top of the current legislation scenario for year 2020 (CLE2020), without any soft mobility measure;
- Scenario 1 in which soft mobility measures are added to the Base-case to substitute 1/3 of the commuters' fleet at the cost of a communication campaign to publicize the measures;

Table 1. Difference in spatial mean of PM10 yearly mean concentrations, costs and savings of SM scenarios compared to the basecase.

Corinair Macro-sector	Δ average PM10 [μg/m3]	Δ AM policy Cost [M€/yr]	Δ AM policy Savings [M€/yr]
Scenario 1	0.02	0.6	352.43
Scenario 2	0.03	126.3	704.87
Scenario 3	0.07	2.0	1079.0

- Scenario 2 in which soft mobility measures are added to the Base-case to substitute 2/3 of the commuters' fleet considering the cost of time spent travelling;
- Scenario 3 that can be considered as a maximum potential scenario, in which soft mobility measures are implemented at their maximum extent (complete shift of commuting traffic to SM), in which the cost for the SM policy adoption is again the communication cost to publicize the measures.

Table 1 presents the differences in average PM10 yearly mean concentrations, costs and savings for the 3 scenarios in comparison with the Base-case. From this Table some facts can be deduced.

In Scenario 3, SM measures are applied at their maximum potential, in other words, all 3 Million commuters adopt one of the 3 different soft mobility measures (walking, cycling or bus travelling). Even applying soft mobility measures at their maximum possible extension only a small improvement in PM_{10} average concentrations can be achieved with respect to the Base-case.

Scenario 1 and 2 assume respectively that 1.15 and 2.30 million commuters adopt soft mobility strategies. Scenario 1 has a lower ambition level, so the SM costs have been assumed again to be equal to communication costs, while in Scenario 2 it has been assumed that, to reach a higher number of active commuters, it is needed to pay them for their increasing commuting time loss. For both these scenarios, half of the active commuters shifting towards SM have been assumed to have abandoned diesel cars and, the other half, gasoline cars. Bus transport is the only alternative for extra-urban commuters, while urban commuters are split between the 3 different transport means.

Scenario 1 has a cost that is similar to the base case (0.66 M€/yr difference), given that it includes only communication costs for AM measures, while Scenario 2 has a significantly higher cost (126.3 M€/yr difference). These scenarios, in terms of PM_{10} concentrations are positioned between the Base-case (no SM measures applied) and Scenario 3, where a complete shift of commuting traffic to SM is implemented.

The high cost increase (124.6 M\€/yr) between Scenario 1 and 2 results in an even higher increase in savings due to reduced fuel consumption (352.44 M€/yr). Meaning that, even if the cost for the application of these strategies are high, the simple fuel savings exceed them.

The particulate matter concentration reduction between Base-case and the feasible scenarios arises mainly from the following precursor emission reductions: 584 ton/yr for NOx and 91 ton/yr for primary PM_{10} in Scenario 1 followed by 1168 ton/yr of NOx and 182 ton/yr of primary PM_{10} in Scenario 2.

The PM_{10} and $PM_{2.5}$ concentrations decrease, reduces the indirect health impact on all-cause mortality due to long term exposure for the whole population. At the same time, the application of SM measures results in health impacts just for commuters. Small variations are estimated among the different scenarios for the health impact due to PM_{10} concentration exposure, with an average per capita YOLL index ranging from 6.21 (Base-case) to 6.18 months (Scenario 3) in a lifetime. YOLL per capita due to increased breathing is a function of the time spent while doing physical activity and $PM_{2.5}$ concentrations and, since the average $PM_{2.5}$ concentrations are not so different, the results are, again, similar between the scenarios and equal to 5.14 and 0.54 months/lifetime respectively for cycling or walking (including commuting by bus). The direct impacts due to increased physical activity, instead, is function only of the time spent while doing an activity and the type of activity considered, so they are the same for all the scenarios and equal to a benefit of 49.7 and 0.24 months/lifetime respectively for cycling or walking.

The application of SM measures imply the reduction of fuel consumption and consequently remarkable savings (from 352M€/year for Scenario 1 to 1079M€/year for scenario 3) and the decrease of GHG emissions. So, the scenario with the lowest reductions (336 kton/year of CO_2eq) is Scenario 1 and the scenarios with the highest reductions are Scenario 2 (249 kton/year of CO_2eq) and Scenario 3 (270 kton/year of CO_2eq).

5 CONCLUSIONS

To efficiently reduce particulate matter concentrations in urbanized environments, especially for sectors where the legislation already provides high levels of application for end-of-pipe measures, may not be an easy task for decision makers. Energy measures and behavioral measures, reducing directly fuel consumption, are a relatively new and not regulated mean to achieve lower pollutant concentrations. This work focuses on a methodology, implemented in the MAQ system, to assess Soft Mobility (SM) measures, including the most relevant impacts in the evaluation process: cost, air quality impact, direct and indirect health impacts, fuel savings and GHG emission reductions.

To evaluate the impact of these measures, a case study on Lombardy region (Northern Italy) has been presented. The SM measures analyzed are applied to commuters and they are: commuting by feet, commuting by bike and commuting by bus.

Three scenarios have been analyzed and compared to a Base-case in which end-of-pipe and energy measures, without considering soft mobility, are applied in a cost-efficient way. These scenarios show that the increasing application of SM measures results always in a PM concentration decrease. Moreover the huge health benefits due to increased physical activity for commuters, the high fuel savings and GHG emission reductions give a clear indication that SM measures can be a significant part of air quality and low carbon win-win policies.

REFERENCES

Arem, H., Moore, S. C., & Patel, A. (2015). Leisure time physical activity and mortality: A detailed pooled analysis of the dose-response relationship. JAMA Internal Medicine, 175(6), 959–967.

Asensio, J., & Matas, A. (2008). Commuters' valuation of travel time variability. Transportation Research Part E: Logistics and Transportation Review, 44(6), 1074–1085.

Carnevale, Claudio, Ferrari, F., Guariso, G., Maffeis, G., Turrini, E., & Volta, M. (2018). Assessing the economic and environmental sustainability of a regional air quality plan. Sustainability (Switzerland), 10(3568).

EEA. (1999). Environmental indicators: Typology and overview Prepared by: Project Managers. European Environment Agency, 25(25), 19.

European Commission. (2005). ExternE Externalities of Energy; Methodology 2005 Update. Reproduction (Vol. EUR 21951).

Guariso, G., Maione, M., & Volta, M. (2016). Environmental Science & Policy A decision framework for Integrated Assessment Modelling of air quality at regional and local scale. Environmental Science and Policy, 65, 3–12.

Guariso, G., & Malvestiti, G. (2017). Assessing the value of systematic cycling in a polluted urban environment. Climate, 5(3).

Mueller, N., et al. (2017). Urban and Transport Planning Related Exposures and Mortality: A Health Impact Assessment for Cities. Environmental Health Perspectives, 125(1), 89–96.

Tainio, M., et al. (2016). Can air pollution negate the health benefits of cycling and walking? Preventive Medicine, 87, 233–236.

Turrini, E., Carnevale, C., Finzi, G., & Volta, M. (2018). A non-linear optimization programming model for air quality planning including co-benefits for GHG emissions. Science of the Total Environment, 621.

WHO. (2013). Review of evidence on health aspects of air pollution – REVIHAAP project: final technical report. Copenhagen: WHO Regional Office for Europe.

WHO. (2014). Health economic assessment tools (HEAT) for walking and for cycling - Methodology and user guide, 2014 update.

WHO. (2017). From Amsterdam to Paris and Beyond : the Transport, Health and Environment Pan European Programme (THE PEP) 2009–2020.

WHO. (2018). Health, environment and climate change - Road map for an enhanced global response to the adverse health effects of air pollution.

Woodcock, J., et al. (2018). Development of the Impacts of Cycling Tool (ICT): A modelling study and web tool for evaluating health and environmental impacts of cycling uptake. PLoS Medicine, 15(7).

Pleasant and attractive public spaces

The public space through an aesthetic ethics

A. Tommasoli
Human Studies Department, European University of Rome, Italy

L. Tommasoli
Department of Urban Planning, Design and Technology of Architecture, Sapienza University of Rome, Italy

ABSTRACT: The contemporary city is based on the interaction between different systems. If we take in consideration the perspective which represents the city as a cultural event (Settis, 2017), the relation, the paper aims to analize, will be between human being and its rapresentation. If we consider the reciprocal influence of the individual and the society, we understand the crisis of the contemporary city as a consequence and a cause, at the same time, of the "crisis of presence". Therefore the contemporary city can be a crucial tool through which developing a reintegration process: take action on the city means make an ethos of transcendence. The paper analyze two case studies aim at redefining the sense of community of urban citizens.

1   BEING-IN THE WORLD

A society is based on the relation between human beings. Exactly, it is based on the relation between a single individual and a plurality of individuals, that is the community. However, on the other hand, such a relation determines the identity of a human being, because it is a continuous process of individualization. It is indeed the determination of a complex and continuous relation between the subject and the society, that the subject expresses through its behaviour. "Agere sequitur esse", writes St. Thomas, explaining the mutual relation in which the essence and the behaviour influence each other, so to make a not-determined subject who constantly works on its identity. Actually, our present western society is determined by an anxiety of belonging and of exclusion from community. It is a cultural crisis that we can consider as the origin of a crisis of the society and of the contemporary city which we call "Crisis of presence". It is a concept created by an italian anthropologist Ernesto De Martino: through the concept of "presence" he means an awareness of society's memories and past experiences. It is indeed the result of a long empirical research that he makes during the 60's in some ethnographic expeditions in the south of Italy, where he finds what he calls "the magic world": a society where the magic exists, because the people believe in it. They act as if it is real and consequently their behaviour belongs to such a magic world. The expeditions of De Martino highlights a particular pathology of this society: the "Tarantism". It is not a phyisical disease infected by the bite of a fields spider (as the population thought), but a psychological matter that comes from the stress of a social exclusion, that is a "Crisis of presence". A social matter that turns into a mental disease and that influences the whole body. The magical idea of a trance is what exactly resolves the social matter: since the subject is affected by Tarantism, it is reintegrated into the society as a sacred human being through some musical rituals. These rituals evoque some traditions, histories, myths that belong to the culture of the community and that strengthen it. This is what De Martino calls "Ethos of transcendence": a metaphysical process of reintegration of the possessed into the magic culture he belongs and a concrete process of reintegration of the single individual into the community he belongs. That is, exactly, a new determination of the individual identity and a new determination of the society in the continuous relation between them. Even if there is not a connection between the anthropology's researches of De Martino and the urban regeneration of the city, the aim we focus on this article is a urban anthropology that works through a philosophical concept that comes from an existential translation of Martin Heidegger's Dasein: being-in-the-world, the specific world of a society, with its culture (traditions, myths, history, beliefs and rituals) that strengthens a community.[1] It is in this perspective that the paper higlights the strategy of urban regeneration as a multi-objective process: recomposing the identity of a community throught the interaction of the social, economic and environmental system. The selected space depute to the re-creation of urban community are the open public spaces in which society have been always represents itself, in order to make an Ethos of trascendence. These spaces are the assigned elements to strenghthen the public image of the city, areas of consensus (Lynch, 1964), encounter between the individual and a common culture. Taking the first step from the urban strategy of the European Commission in the 80's, (Urban Pilot Project, 1989-1994, 1997-1999; URBAN I 1994-1999, URBAN II, 2000-2006) (De Santiago Rodríguez, 2017), scientific international debate has been defining different goals between which, one, is the most connected to the Ethos of Trascendence:"make cities and human settlements inclusive, safe, resilient and sustainable" (UN, 2015).

2 GUIDE LINES AS A MAIN GOAL

There is a polarity between the morphology and the symbolism of a place. The form of an open public space, like a square, is universally recognizable both as a framework to group house in a settlement, than as a courtyard with a symbolic value for inhabitants: "we might almost infer the existence of a kind of social ritual, which produces a perfect match between individual and collective"(Krier, 1979). The course of the market had led to a continue meta-morphosis of society. Contemporary cities perform these changes through a loss of symbol-ism: the lack of open public spaces represents today the crisis of presence. The main goal of the paper is to define guide lines to guarantee open public spaces, which, through morpho-logical, structural and functional interventions, can regain human being consciousness of the sense of community, executing the ritual Presence of human being.

3 HOW TO LOOK FOR GUIDE LINES IN OPEN PUBLIC SPACE

Through an inductive method, the paper analyzes two case studies, different both in scale than in the object of the analysis, in order to define strategical guide lines for a network of public spaces dissemenated in the urban territory capable of recreating the rituality of the community. The two cases are placed in fluvial spaces: natural components of landscape which passing through the city. They are caesur and at the same time a continuity: border. The morpholgy of being-in-presence can be rapresented by the limit or the border. Two different concepts which represent the phenomenological transposition of both phisical than conceptual space. The limit is what prevent by an action, which can be a moviment of a person to another (being in presence with), or a prevention from a moviment in a space (being in presence in). The border is the delimitation of a line that both phisically than metaphorically represents a continuity of moviment with the con-sciuscness of a detachment. The phisical border is the definition of a space which can be crossed or of a space which can contain, taking in consideration it as an instruments of urban policy and strategy. We indeed can state the border as an advantageus element: it enriches the human being with a new knowledge, elevating space with a new meaning. Since it's origin Civility used to settle itself on the border between soil and water, near the limit of a see side front or a river front.

4 BETWEEN CITY AND WATER IN LISBON

The case "Devolver o Tejo às Pessoas, intervenções na Frente Ribeirinha – Baixa" led us enter into the main focus of the urban regeneration: people. It is clear from the plan that this part

---

1. *Heidegger, M, Building Dwelling Thinking, from Poetry, Language, Thought, translated by Albert Hofstadter, Harper Colophon Books, New York, 1971*

Figure 1. The functional programme of the urban regeneration in Lisbon.

of the city is fragmantated in tree different settings: in the middle part there is a rinascimental planned, on the east side the ancient city, and on the opposite side there is an area in-between, designed according with the rinascimental axises togheter with ancient tissues. The morphological discontinuity is resolved throught out the intervention: a functional program setteld along the river as a connection with its articulation between the west riverside area, the hill of "San Francisco" and the urban area "Nascente da Ribeira das naus" and, at the same time, as the main interface of public transport: inland waterways, railways, buses, trams and subways (Figure 1). The first fase of the project, finished in 2013, named "Requalificação do espaço público cais do sodré/corpo sant", aims at valuing the use of public space by the pedestrian through the extension pestrian paths, namely the riverside walk, recovery of the typology of the Square-Garden, "reconstruction" the Largo do Corpo Santo, increased afforestation and renovation of urban furniture in the new areas of residence. The project answer to the will of "preserve, reuse and enhance the pre-existing distinctive elements of this space"[2].

4.1 Space of awareness

The main focus of the intervention is the open public space, as a fluid involving all the different function in a continue path. The open public space represents "the space of the awareness": citizens can get back to represents themselves as an ancient urban ritual. An urban regeneration with its integrated strategy purposes a solution for the different systems: the connective, the recreational, the cultural and ritual ones. The rite that was once considered daily interrelationship between man and water as a tool of work and sustenance, now it opens the possibility of new relational definitions.

5 BETWEEN CITY AND WATER IN ROME

The case of Rome is about the management of a part of the city which is depute, according to the general development plan of Rome of 2008, to be a strategical area[3]. The paper aims to introduce a specific intervention on a part of the river organized by the association Tevereterno Onlus, which creates the event "PiazzaTevere". The main goal is to give back to citizens the river as an open public space, throught cultural events. Focused more on the management of the area this case is not properly a case of urban regeneration but a case of cultural intervention in the city, composed by a startegical functional programme.

5.1 Mythology as ritual

"Triumphs and Laments" is a site-specific artistic work, by the South African artist William Kentridge, 500 meters long, representing about 80 figures, up to 10 meters high, that give life to a procession of silhouettes along the Tiber, in the stretch between Ponte Sisto and Ponte

---

2. The project is available at: http://tevereterno.org/progetti/triumphs-and-laments/
3. Ambiti di Trasfromazione Startegica referring to: http://www.urbanistica.comune.roma.it/prg-adottato/prg-adottato-elaborati-descrittivi/prg-adottato-d7.html

Figure 2.   Triumphs and Laments, Tevere, Rome.

Mazzini (Figure 2). The figures represent reinterpretations of the artist of an iconographic selection of characters and historical facts that have marked the millenary history of Rome from the classical and mythical age to the modern events of the most recent past. It was inaugurated with an extraordinary musical and theatrical free event, conceived in collaboration with Philip Miller that superimposed the figures realized on the wall other figures created with an infinite play of Chinese shadows. This event has brought back to memory the ancient relationship that once used to connect the inhabitants of Rome with their river: a sacred place. Untill the '60 the city used to act a ritual event each year: a wedding between a princess (usually a famous italian woman) and the river. The urban tissue as a cognitive experience:it organizes what we see on the basis of imaginative, emotive, identifying and memorial components, revealing themselves through a satisfaction or dissatisfaction that leads to the recognition of the value or disvalue of what we see (D'Angelo, 2010).

6   CONCLUSIONS

The crisis of presence is strictly connected to the crisis of the contemporary city because of the reciprocal influence between the individual and the society. Therefore we must think about a urban regeneration as an ethos of transcendence: a reintegration of the individual in the society through a cultural process that is a ritual in which live the community's traditions, myths, history and beliefs. Thus we highlighted the urban regeneration as this cultural process recomposes the identity of a community and of the individual. It means that the re-creation of the urban community through the re-generation of the public spaces is an ethical work. In this way, the case studies of Lisbon and Rome helped us to understand why and how can be succesfully realized such an anthropological process of urban regeneration in the contemporary cities. The cited UN goal is a cornerstone that higlights the global necessity to resolve the urban problem of inclusion. An inclusive city can face the anxiety of the individual of being excluded, an anxiety which affects society itself, leading, nowdays, to conflict and violent urban situacions. Towards a return to the city as a common good par excellence, the human being can face the crisis of presence through spaces in which executing the ritual of being-in-the-world: creating relationships.

REFERENCES

Camara Municipal de Lisboa, Uma cidade para as pessoas, Lisboa.
De Martino, E., The Land of Remorse: A Study of Southern Italian Tarantism, translated by Dorothy L, Zinn. London; Free Association Books. 2005.
De Martino, E., Primitive Magic; the Psychic Powers of Shamans and Sorcerers, Bridport: Prism Press, 1988.
D'Angelo, P., (2010), Filosofia del Paesaggio, Quodlibet, pag.23
Heidegger, M, Building Dwelling Thinking, from Poetry, Language, Thought, translated by Albert Hofstadter, Harper Colophon Books, New York, 1971.
Lynch, K., (1964). Title Title Title Title Title Title. Editor. Pages.
Krier, R.,(1979), Urban Space, Rizzoli, pag.15
Settis, S., (2017), Architettura e Democrazia, Paesaggio, città, diritti civili, Einaudi Editore.

Pedestrians, Urban Spaces and Health – Tira, Pezzagno & Richiedei (eds)
© 2020 Taylor & Francis Group, London, ISBN 978-0-367-46171-3

Imageability of geo-mining heritage. Case study of Nebida settlement, Geo-mining park Sardinia (Italy)

N. Beretić & T. Congiu
Department of Architecture, Design and Urban Planning, University of Sassari, Alghero Italy

ABSTRACT: With respect to the scale of observation and the low-urbanized context, this paper argues perceptual qualities that help the interpretation of heritage settings experienced by walking. Focused on physical environmental indicators and aesthetic qualities to be used in Geo-mining heritage settings, the paper redefined qualities of visual impact from previous findings including geometric features of a path (profile and edges) and imageability as comprehensive perceptual quality. Selected indicators were implemented at two selected routes in the ex-mining settlement of Nebida. Results demonstrated relationships between indicators of path geometry and of imageability; the more complex geometry, the more diverse imageability and experience of walking. Moreover, in the case of Nebida, planned and maintained route offers the less direct and variable experience of Geo-mining heritage than informal track beaten by hikers.

1 INTRODUCTION

A review of transport & land use research and public-health literature reveals a range of objective variables used to analyze and characterize the built environment which differ in measurement methods and rough data (Choi, 2012; Vale et al. 2016; Blecic et al. 2017). Some functional factors are standard and applicable on all scales and contexts while other features are more variegate, prevalently those related to the condition and aesthetic qualities of urban design. Less investigation has been conducted at the lanscape scale finding easthetics as the most important factor in promoting walking in rural contexts (Li et al. 2018).

In Geo-mining Parks, heritage elements scattered in the landscape often away from main settlements, are inseparable from the natural landscape scenery (Beretić et al. 2019). Dealing with the perception of such landscapes is a matter of high importance for their valorisation and regeneration; yet, it is also a matter of high complexity (Misthos et al. 2016). Thus, this paper argue the fundamental need to combine values provided by aestetic qualities of two different spatial languages (distinctive of the active transport field and landscape architecture). The paper aims to redefine the 'perception' influencing factors and indicators of the imageability of the phisical environment of Geo-mining Parks. We implemented the redefined qualities in the former mining settlement of Nebida (south-west Sardinia, Italy) selected as case study.

2 STUDY AREA AND METHODS

Nebida is a costal settlement emerged for the purposes of intensive mineral extraction (Mezzolani et al. 2007). Located in the area of Iglesiente, Nebida is a segment of the Geo-mining Park that has been recognized as a heritage of great importance by UNESCO in 1997. Figure 1 shows the settlement of Nebida in the general context and position of two routes selected for this study.

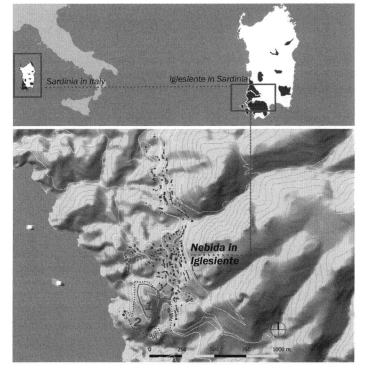

Figure 1. The general context of Nebida settlement and two selected routes.

A multi-disciplinary theoretical framework guided this research. We used the case study research method (Yin, 1994; Flyvbjerg, 2006), including direct observations (Taylor-Powell et al. 1996) for empirical evidence.

The conceptual development of the research consists of the following steps: a) desk and field data collection (literature about perception and Nebida settlement); b) analysis (selection of the qualities and features that affect walking and landscape perception); c) first set of results (definition of the qualities and features of landscape perception congruent with active mobility); d) second set of results (application of the defined criteria at the case study of Nebida and discussion of empiric results).

3 DESCRIPTION

3.1 *Research development and theoretical results*

Departing qualities of visual impact considered, were geometric features of path (e.g. path profile and route anathomy) and five perceptual qualities (imageability; enclosure, human scale, transparency and complexity) operationalised in previous studies (Ewing et al. 2005; 2009). Successively, to obtain indicators suitable for analysing Geo-mining heritage settings, we compared the above-mentioned qualities with those from the research of landscape as three-dimensional compositions in rather natural context: stewardship, coherence, disturbance, historicity, visual scale, imageability, complexity, naturalness and ephemera (Tveit et al. 2006). Despite this comparison revealed differences in the approaches and indicators divergences, the concept of imageability refers most to common background theories (Lynch, 1960; Rapoport, 1987) and incorporates some of the other qualities. We focused on this quality and possibilities to operationalise features objectively and easily identifiable. We defined imageability as unique, distinctive and memorable qualities of a landscape impression present in totality or through elements. Elements, both natural and cultural, making the landscape create a strong visual image in the observer. It is one whose elements are easily identifiable and grouped into an overall pattern. Imageability relates many other spatial qualities. Exclusive to this paper, imageability describes properties of the physical landscape itself. Potential indicators of imageability are grouped in tow categories. First, indicators of general aesthetic and perception of a landscape image include viewpoints, presence of historic and cultural features heritage elements) and patterns, major landscape features, presence of spectacular, unique or iconic elements and landmarks, presence of water bodies and moving water, and viewshed size and form. Secondly, indicators of path profile (twist and turn, direction/width changes, asymmetries and elevation gradients) and pathe edges (depth of vision, frontages (dis)continuity, contrast of surfaces, enclosed spaces and equipments).

4 IMAGEABILITY AND PERCEPTION OF HERITAGE ELEMENTS IN THE NEBIDA SETTLEMENT

We implementd the proposed analytical method into two paths that run through the Geo-mining landscape of Nebida (Figure 2) and differ for their structural characteristics and

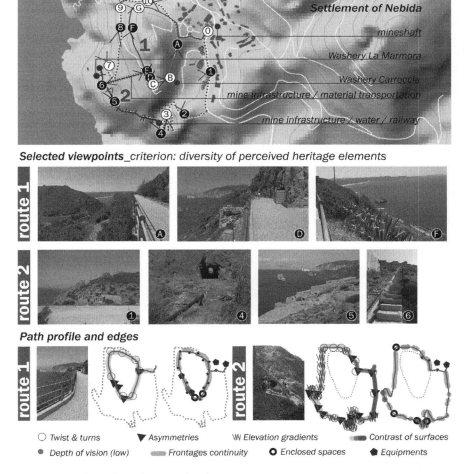

Figure 2.   Selected viewpoints of two analysed routes.

experience of the place. First they have diverse length, profile and degree of accessibility: Route 1 is planned, equipped and regularly maintained while Route 2 is a longer informal track beaten by hikers. An almost constant view on the sea occurs along both paths. R1 is a short panoramic loop around the hill, accessible to all because of its flat profile, paved surface and presence of railing and equipments (benches, small funtains, repair elements and a restaurant). It offers a wide hill-top view of the Geo-mining site with close-up views on heritage and natural environment. R2 is a long winding track, more difficult to walk for its steepness and uneven surface. Running alongside the hill with twist and turn sequences it allows to reach and visit the abandoned mining structures located on the hillside arousing an exciting experience of the heritage landscape.

To determine the major viewpoints we combined structural singularities of the path (asymmetries, direction and elevation changes, discontinuity of frontages, depth..) with sharp vision and perception of both heritage elements and unique iconic elements and landmarks. Selected viewpoints represented in Figure 2 are positioned before the occurrence of important heritage elements and landmarks whose succession along the path supports the understanding of the former fucntionalities in the Geo-mining process. In such a way, while walking observers can retrace the original pathways of materials and workers realizing how the Geo-mining site worked.

Due to their geometry and position the two paths entail different relationships with the heritage elements and as such different experience of walkers. While R1 emphasises the scenery, giving a view of the Geo-mining elements from afar and focusing on their integration in the natural setting, R2 enters and traverses the landscape revealing gradually its components and bringing visitors in direct contact with heritage features that can be experienced by all senses and not exclusively by visual perception. Nevertheless, if the deeper immersion of R2 offers a more exciting experience it requires walkers to be trained and in good health conditions. On the contrary R1 presents less variations but can be walked comfortably by everyone, including disabled thanks to its regular surface and furnitures.

5 CONCLUSIONS

The paper offered a set of imageability influencing indicators representative of the aesthetic qualities and features of the phisical environment that affect the experience of walking in Geo-mining Parks, encourage active mobility and enhance the perception of Geo-mining landscape. A qualitative and operational definitions is given making use of two distinct spatial languages: one pertaining to the active transport field and the other distinctive of landscape architecture.

We made use of metric reality for its more objective nature in comparison to measurements of soft landscapes and detail patterns (touch, smell, textue or time) although equally fundamental in landscape composition.

The "specific and extreme character" (Sklenicka, 2010:424), of mining/post-mining landscapes makes this research an exceptional case when compared to the more common cases of agricultural or urban landscapes (Misthos et al. 2016). This condition could be considered a limit in the replicability of the method. However, the lack of literature about active mobility and perception in the rural and Geo-mining Parks context, the prevalence of qualitative research regarding the perception of the mining landscape, and the attention observed in this research for imageability influencing factors easy to collect and objectively measurable make it a pioneer and specialized study on the topic of the relations between perceptual qualities of landscape settings and recreational walking. With this purpose further investigation is needed, with two aspects considered overriding: the quantification of selected indicators and the inclusion of landscape patterns in the analysis of imageability.

ACKNOWLEDGEMENT

This research is conducted and funded within the Tsulky research project. The Tsulky research project ("Tourism and Sustainability in Sulcis-Iglesiente area") was funded by Autonomous

Region of Sardinia (within the Agreement for the management of fundamental or basic research projects for the implementation of interventions under the research for the "Sulcis Plan", CIPE Resolution n. 31 20.02.2015, and resolution n. 52/36 28.10.2015, "Sulcis Strategic project" – public-private research project), executive resolution n. 6525 register n. 666 14.09.2017. The APC was funded by the project chief scientist for the 3rd research unit, prof. Alessandro Plaisant.

REFERENCES

Beretić, N. and Plaisant, A. (2019). Setting the Methodological Framework for Accessibility in Geo-Mining Heritage Settings: An Ongoing Study of Iglesiente Area (Sardinia, Italy). Sustainability 11(13):3556.

Blecic, I., Canu, D., Cecchini, A., Congiu, T. and Francello, G. (2017). Walkability and Street Intersections in Rural-Urban Fringes: A Decision Aiding Evaluation Procedure. Sustainability, 9:883.

Choi, E. (2012).Walkability as an Urban Design Problem. Understanding the Activity of Walking in the Urban Environment. Ph.D. Thesis, KTH Royal Institute of Technology Architecture and the Built Environment School of Architecture, Stockholm.

Ewing, R. and Handy, S. (2009). Measuring the Unmeasurable: Urban Design Qualities Related to Walkability. Journal of Urban Design. 14:65–84.

Ewing, R., Clemente, O., Handy, S., Brownson, R.C. and Winston, E. (2005). Identifying and Measuring Urban Design Qualities Related to Walkability. Final Report prepared for the Active Living Research Program of the Robert Wood Johnson Foundation.

Flyvbjerg, B. (2006). Five Misunderstandings about Case-Study Research. Qualitative Inquiry. 12:219–245.

Forsyth, A., and M. Southworth (2008). Guest editorial: Cities afoot—pedestrians, walkability and urban design. Journal of Urban Design 13:1–3.

Lynch, K. (1960) The Image of the City. Cambridge, MA: Joint Center for Urban Studies.

Mezzolani, S. Simoncini, A. (2007). L'archeologia industriale. In Mezzolani, S., Simoncini, A., (Eds.), Sardegna da Salvare. Storia, Paesaggi, Architetture delle Miniere, 3rd ed. Nuoro, Itlaly, Archivio Fotografico Sardo, p. 11–20.

Misthos, L.M. and Menegaki, M. (2016). Identifying Vistas of Increased Visual Impact in Mining Landscapes. Proceedings of the 6th International Conference on Computer Applications in the Minerals Industries - CAMI 2016, October 2016. Istanbul, Turkey.

Rapoport, A. (1987). Pedestrian Street Use: Culture and Perception. In A. Moudon, Public Streets for Public Use. New York: Van Nostrand Reinhold Company Inc.

Sklenicka, P. and Molnarova, K. (2010). Visual perception of habitats adopted for post-mining landscape rehabilitation, Environmental management 46(3):424–435.

Taylor-Powell, E. and Steele, S. (1996). Collecting evaluation data: Direct observation. In Program Development and Evaluation. Madison, WI (USA), University of Wisconsin-Extension p. 1–7.

Tveit, M., Ode, A. and Fry, G. (2006). Key concepts in a framework for analysing visual landscape character in Landscape Research. Routledge. 31(3):229–255.

Vale, D.S., Saraiva, M. and Pereira, M. (2016). Active accessibility: A review of operational measures of walking and cycling accessibility. J. Transp. Land Use; 9:1–27.

Yin, R. (1994). Case Study Research: Design and Methods, 2nd ed. Beverly Hills, USA, Sage Publishing.

Li, C., Chi, G., & Jackson, R. (2018). Neighbourhood built environment and walking behaviours: evidence from the rural American South. Indoor + built environment: the journal of the International Society of the Built Environment, 27(7):938–952.

Owen N, Humpel N, Leslie E, Bauman A, Sallis JF. (2004). Understanding environmental influences on walking: review and research agenda. Am J Prev Med. 27(1):67–76.

Pedestrian healthcare and beauty: Free-accessibility design plan in Taranto (IT)

A. Massaro
MSc Uniparma, Taranto, Italy

F. Rotondo
Università Politecnica delle Marche, Italy

ABSTRACT: This paper aims to discover if and how the Free-Accessibility Design Plan in Taranto, can improve the rediscovery of local identity and even if historical identity together with economic criteria can help to reinforce urban pedestrian accessibility. Local identity can be used as a proxy of urban beauty (at least it's historical beauty). In this case, urban beauty is directly connected to identity and costs. Thus, the Free-Accessibility Plan becomes a relevant part of sustainable urban mobility strategy which affects the land use decisions, within an overall urban regeneration policy. So, the paper illustrates the structure of the plan, its methodology and results inside the urban regeneration strategy.

1 A BRIEF INTRODUCTION

In the urban planning process nowadays ongoing in Taranto city, the Free-Accessibility Design Plan (FADP) has a transversal role, depicted by the different actions that previews, in which it can be considered involved by the higher-level planning. In this search, we manage the GIS network analysis to understand the capability to acquire more economic and time benefits putting up beauty in the design plan. The beauty of urban space is evaluated focusing on identity: three different parts of the city in the same city-centre, each one with its proper character and its proper identity, and the relative main flows of pedestrian connections to check the structure of the most used urban network. The evaluation of the cost to improve walkability in the main paths strengths the urban structure development, so the accessibility and the serviceability of the spaces, the cost-effectiveness of the urban identity.

2 URBAN PLANNING AND MOBILITY IN TARANTO

If the current Master Plan is superimposed on a recent orthophoto (Figure 1), it becomes clear that the city has evolved independently of the plan, in recent decades. Master Plan's variations have been numerous and the presence of "spontaneous" buildings spread mainly along the coast within the strip located at a distance of 300 m from the sea, is equally wide.

It can be seen that on all the green and blue areas, which the urban plan has destined to services and public utilities, there are residences, often arisen abusively.

This unplanned expansion has certainly made even more complicated the mobility of a city of 200,000 inhabitants born on an island and developed around the two seas that distinguish it, but that make it not very accessible to cars and public transport on road or on iron (the only alternative suited to the geographical features is the waterway). The question of the environmental or social costs of urban form is increasingly attracting attention in spatial policy (Camagni, Gibelli, Rigamonti, 2002; Jia, Tang, Xu, Yang, 2019) and the case of Taranto is highly interesting in this field.

Figure 1. Extract of the general town planning of the city (Source: www.comune.taranto.it).

To this situation determined by the geography of the places and by the bad habits of the local communities, is added an inability to plan and complete the numerous urban planning paths started and never concluded after the approval of the current General Town Planning (Piano Regolatore Generale of the 1978).

The current Administration seems to want to change this situation, having started the new Master Plan and approving the Urban Sustainable Mobility Plan (SUMP). As put in evidence by Banister (2008), conditions for change are dependent upon high-quality implementation of innovative schemes, and the need to gain public confidence and acceptability to support these measures through active involvement and action. That's why, both plans have been put in practice with high inhabitants' and stakeholders' involvement.

This is the framework in which we will try to outline the ionic system of mobility and logistics, observed not as a mere planning of the sector, but as an essential element of the general urban planning of the city and its future.

2.1 The mobility and logistics system: Ongoing policies and expected results

What emerges from a very concise picture is that the events that characterize the history of Taranto are, in most cases, linked to choices not directly made, but often "suffered" by local actors, linked in the first place to the geographical position and the physical characteristics of the city.

Thinking of the port of Taranto, which, in a strategic position with respect to the main routes between East and West, represents a strong point of the city, together with the dense infrastructure network, of the "sky" (Grottaglie Airport), of the "iron" (served by the state and south-east railways) and the "land" (end point of the A14 motorway) which give it a privileged position.

Going down the scale, on an urban level, the city has chosen to plan urban planning and mobility through two different but coordinated and coherent instruments.

The Preliminary Programmatic Document (DPP) which represents a first proposal of the new General Urban Plan (PUG) and the Urban Sustainable Mobility Plan (SUMP) which has declined the overall strategy of urban mobility in coherence with the possible urban scenarios outlined in the DPP.

Consistent with the European guidelines (2013) and European good practices (May, 2015), the SUMP has chosen a possible solution of the ionic transport system oriented to a simpler modality of implementation and effectiveness in relatively short times and possibly with reduced impacts on the environment, trying to pursue multimodal sustainable mobility in urban areas.

He identified two main problems of ionian mobility: the lack of inter-modality and the overwhelming weight of the traffic crossing the "Borgo" quarter and the "Old City" quarter which congested the city centre and did not allow different uses, damaging also the urban trade that in these years is seen to be overtaken by large shopping centres.

In this strategy, pedestrian and bicycle mobility play a central role.

Leaving aside the basic concept that permeates the entire SUMP (pursuing multimodal sustainable mobility), the Plan operates through different phases trying to investigate (through computer models) what would happen if a Limited Traffic Crossing Zone (ZTAL) was introduced which try to decrease and then eliminate the traffic crossing from "Borgo" and "Città Vecchia" quarters.

In the second phase, the strengthening of Local Public Transport is proposed through the reorganization of accessibility to the railway system with the reactivation of the "Nasisi" railwaystation and the strengthening of urban transport with the creation of a "supporting" network based on two lines of Bus Rapid Transport (BRT).

The BRT differs from a conventional bus line for the range of vehicles (generally articulated to guarantee adequate capacity), the preparation of the stops (flushing and facilitating the approach of the vehicle to the pavement in order to facilitate passengers with temporary or permanent reduced mobility), the reserved way and the traffic light preference at the stops (Prayogi, 2016; Breipthaupt et alii, 2014).

This action is accompanied by the identification of a series of interchange parking lots (park and ride).

In the SUMP the priority must be assigned to the restitution of a "beautiful" and inclusive image of the city by acting on the redevelopment, regeneration and recovery of the building heritage and the urban space, which are the main prerequisites for inducing a modal diversion from the private car to active mobility (pedestrian and cycling).

The possible pedestrianization of part of the Old City and the increase of pedestrian areas in the "Borgo" quarter, together with the introduction of cycling routes throughout the city, will favour the affirmation of these additional sustainable modes of transport.

The SUMP was followed by the adoption of the Plan for the elimination of architectural barriers which, by improving the accessibility and safety of pedestrians, consolidates the potential use of pedestrian paths.

3 METHODOLOGICAL APPROACH

According to the European Agenda (Urban Agenda for the EU, 2016), this reserch underlines the role of "accessibility (for disabled, elderly, young children, etc.)" in the planning system, promoting "the equal rights of all persons with disabilities to live in the community, with choices equal to others" (Article 19 of the CRPD[1]). By this way, the FADP develops its phases, triing to investigate the issues regarding the physical characteristics of the pedestrian network, examined by the construction of a GIS Network System, explaining the complexity of the pedestrian flows and its proprieties (Giulio Maternini, 2017), (Paolo Ventura, 2017), (XU, 2014). The multiplicity of constrictions that affects the paths is put in evidence, above all for the structural paths, chosen by the analysis of the interactions with the future SUMP. In this way, it is possible to subdivide the planning zone in different microzones of priority by the different indicators of safeness and by using the frequency of flows.

---

1. "Convention on the Rights for Persons with Disabilities" (Disabilities, s.d.)

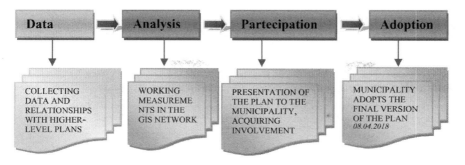

Figure 2. Phases of the free-accessibility mobility plan.

The built network bases its proper capacities on the linkages evaluation between the different nodes.

In this research about the Taranto's pathways system, evolution has had different meanings, which have been married to the different indicators of the made analysis. The specific state of the art of every path gets us informations about the historical use of these. Acquiring step by step a different role in the metabolism of urban mobility development, the most relevant public buildings gain an important node-role in the network. Specifically, they help giving identity to each one of the three different parts of the city analyzed. The focus on historical public buildings helpful in an effort to comprehension the most intimate weave of connections of the city, so its beauty, expression of the chapters of its history, its unicity.

Specific indicators (Safety, Health, Wearing, ect.) have been associated to every path to evaluate the cost of the renovation. Renovation addicts to the need for accessibility, regarding the most affected micro zones of the plan by constrictions. So discovering the best linking paths, managing the different depicted indicators, it is possible to give a cost to the renovation for each one of those. To trigger the regeneration process, time acquires a main role: reduction of the journey time implementing accessibility to the shortest path, it means to guarantee walkability and stresses the development needs of the surroundings (Figure 2).

Figure 3. Structural pedestrian paths on overlapping map.
(Source: Authors).

Structural pedestrian paths subdivided by cost of interventions, normalized by time path in six black classess by thickness of black line. Those are overlapped by the service area related to the time journey (2, 5, 10, 15 minutes) to link inhabitants to the main buildings of the area (the buildings are depicted by their entrahances, subdivided in accessible and inaccessible). Moreover, to the map it been overlapped the densify map of the GDPpi normalized by civic number. It is enough to understand the value of re-sewing and interconnection played by the chosen structural paths. By the way, the yellow line underpins the main path of the entire planning area.

3 CONCLUSIONS

There are positive effects deriving from accessibility implementation, so Free-Accessibility Design Plan should be connected to the planning system as a structural operative plan for mobility, so for city urban planning.

In the end, in a quick perspective, the Free-Accessibility Design Plan would help the "real change on the ground to ensure people with disabilities can enjoy their rights in practice" (according to the FRA Director, Morten Kjaerum (Rights, 2016)).

The cost of the investments is depicted for the structural paths (Figure 3), finding strength connection between the FAMP and the PUMS, where the first one really becomes an operative tool of the second one. Moreover, the made evaluation on the healthcare system and the beauty of the area is an effort for the PUG's perspective, managing the walkability in the strenght lines of the urban identity. In this interaction among the different parts of the Taranto urban planning, it is possible to underline a common denominator in the accessibility, describing a new structured and organic system ongoing.

REFERENCES

Banister, D. (2008) The sustainable mobility paradigm. Transport policy, 15, pp.73–80;

Breipthaupt M., Martins W. C., Custodia P., Hook W., McCaul C.(2014), The BRT Standard, New York, Institute for Transportation and Development Policy.

Camagni R., Gibelli M. C., Rigamonti P., (2002), Urban mobility and urban form: the social and environmental costs of different patterns of urban expansion, Ecological Economics 40/2, pp. 199–216;

Disabilities, C. o. (2016). https://www.un.org/. Tratto da https://www.un.org/development/desa/disabilities/convention-on-the-rights-of-persons-with-disabilities.html.

European Union, (2013), Guidelines. Developing and Implementing a Sustainable Urban Mobility Plan, available on line at: https://www.eltis.org/sites/default/files/guidelines-developing-and-implementing-a-sump_final_web_jan2014b.pdf, retrieved on the 16th of august 2019.

Ferbrache F., (ed., 2019), Developing Bus Rapid Transit. The Value of BRT in Urban Spaces Transport, Mobilities and Spatial Change.

Giulio Maternini, F. G. (2017). Progettazione e tecniche di itinerari e attraversamenti. Forlì: EGAF edizioni;

Jia, Y., Tang, L., Xu, M., Yang, X. (2019), Landscape pattern indices for evaluating urban spatial morphology – A case study of Chinese cities, Ecological Indicators, 99, pp. 27–37;

May A. D., (2015), Encouraging good practice in the development of Sustainable Urban Mobility Plans, Case Studies on Transport Policy, Vol. 3, Iss. 1, pp. 3–11;

Paolo Ventura, M. Z. (2017). Gis-based monitoring and evaluation system as an urban planning tool to enhance the quality of pedestrian mobility in Parma. Parma: Department of Engineering and Architecture, University of Parma, Italy. Tratto da www.researchgate.it.

Prayogi L., (2016), The Influence of Bus Rapid Transit System on Urban Development: An Inquiry to Boston and Seoul BRT Systems' Performance Indicators International, Journal of Built Environment and Scientific Research 1 (1), 1–7;

Sagaris, L., Tiznado-Aitken, I., Steiniger, S., (2017), Exploring the social and spatial potential of an intermodal approach to transport planning, International Journal of Sustainable Transportation, 11, pp. 721–736; Steering the Metropolis: Metropolitan Governance for Sustainable Urban Development, pp. 225 – 238;

Travisi C. M., Camagni R., Nijkamp P., (2010), Impacts of urban sprawl and commuting: a modelling study for Italy, Journal of Transport Geography, 18, pp 382–392;

Urban Agenda for the EU, P. o. (2016). https://ec.europa.eu/. Tratto da https://ec.europa.eu/regional_po licy/sources/policy/themes/urban-development/agenda/pact-of-amsterdam.pdf.

XU, M. (2014). A gis-based pedestrian network model for assessment of spatial accessibility equity and improvement prioritization and its application to the Spokane public transit benefit area. (M. XU, A cura di) Tratto da www.researchgate.it.

Zegras C., (2017), Metropolitan Governance for Sustainable Mobility, In: Gómez-Álvarez D., Rajack R., López-Moreno D. and Lanfranchi G., Inter-American Development Bank, United Nations, Human Sttlements Programme and Development Bank of Latin America, Steering the Metropolis: Metropolitan Governance for Sustainable Urban Development, pp. 225–238.

Pedestrians, Urban Spaces and Health – Tira, Pezzagno & Richiedei (eds)
© 2020 Taylor & Francis Group, London, ISBN 978-0-367-46171-3

The effect of Movida on residential property prices: An example from Turin

E. Ottoz
Università di Torino, Italy

P. Pavese
Università del Piemonte Orientale, Italy

L. Sella
CNR-IRCRE, Italy

ABSTRACT: European cities are experiencing a particular type of noise pollution originated by nighttime recreational activities, the so called "Movida", which hasn't been properly investigated yet. The aim of the paper is to examine the effect of recreational noise on residential property prices. We used an original highly detailed housing transactions dataset from the City of Turin covering the period 2017 to 2018 and built an indicator of recreational noise based on the proximity of dwellings to the night recreational activities. The results obtained employing hedonic modelling show that the adverse environment for an apartment located in a "Movida" district will result in a lower market value as compared to an apartment with similar characteristics, except for recreational noise.

1 INTRODUCTION

According to the World Health Organization (WHO-JRC, 2011): "..noise pollution is considered not only an environmental nuisance but also a threat to public health." The most common sources of noise pollution, and consequently those which have been mainly investigated, are related to traffic and industrial activities. However, in the last three decades European cities have been affected by a particular type of noise pollution stemming from night recreational activities generally located in the city centers, the so-called movida.

The approach to nigh- time economy is complex, as the phenomenon carries high potentials in terms of social and economic benefits, but also problems related to the impact of alcohol on crime and disorders, coupled with public nuisance caused by recreational noise pollution, originated by crowds in the streets at late hours and loud music. See Bevan (2011), Hadfield (2017) and Ottoz et al. (2018).

In spite of its relevance and diffusion, the implications, both social and economics, have been poorly investigated so far.

2 OBJECTIVE OF THE PAPER

Hedonic regression technique has often been used in the literature to study the effect of road, railway and airport noise on property prices, but little evidence exists on the effect of recreational night time noise: the aim of the paper is then to examine the role played by activities such as pubs, discos, restaurants, on residential property prices

We used an original highly detailed housing transactions dataset referring to the City of Turin covering the period 2017 to 2018 and built and indicator of recreational noise based on the proximity of residential properties contained in the data set to the recreational activities operating at night. A property represents not only an amount of structural characteristics, but also a set of specific locational characteristics. Next section briefly reviews the empirical literature on hedonic pricing and noise section 3 describes the dataset. Section 4 presents the main estimation results. Section 6 concludes.

3 RESEARCH APPROACH

3.1 *Data*

The empirical analysis in this paper is based on two main datasets.

The first dataset contains the exact geo-referenced localization of recreational activities in the central districts of the City of Turin and other related information, including opening days and closing hours for a large part of commercial activities. It has been extracted by Google Maps Places API, which returns information on global points of interests, including bars, nightclubs and cafes, which are the main drivers of movida nighttime noise.

The second dataset covers about 60% of all housing transactions in the central areas of Turin (Centro, Quadrilatero, Vanchiglia, San Salvario) from 2017–2018 and is obtained from the Osservatorio del Mercato Immobiliare of Agenzia delle Entrate and Italian Association of Realtors (AICI) and microdata from each transaction.

The variables included in the regression are:

- *Dimension*: size of the property in squared meters (sqm).
- *Physical building characteristics*: age, size, and number of bedrooms, bathrooms.
- *Walkability*: the distance of the apartment to relevant points of interest (POI) as schools, retail, recreational, entertainment and food destinations within 250 meters.
- *Accessibility*: a set of dummy variables, reflecting the opportunities that retail properties are easily reached by all transport modes.
- *Movida*: recreational noise proxies based on the proximity of dwellings to "Movida" commercial activities.

Tables 1 reports the descriptive statistics for the sample of transactions in the central districts, 1793 observations.

3.2 *Empirical results*

The hedonic function we estimate is the following:

$$\ln P_i = \alpha + \sum_{k=1}^{k} \beta_k z_{ki} + \varphi noise_i + u_i \tag{1}$$

where *lnP$_i$* is the logarithm of the price per square meter of dwelling i; β$_k$ is the implicit price of the k-th characteristic, z_k is a vector of characteristics; *noise$_i$* is the recreational noise proxy with implicit price φ; u_i is the random error term. Note that the social valuation of house characteristics does not change over time.

We use semi-log specification (denominated log-lin), which is the most widely used functional specification because it helps to normalize the price and residual distribution, allowing the results from different studies to be compared. Then, the coefficients of the estimated hedonic equation can be interpreted as semi-elasticities, all the other characteristic the same.

Table 2 reports the estimated results from the semi-log specification. According to standard criteria regarding the goodness-of-fit, our model explains a large share of the variance

Table 1. Descriptive statistics (n = 1,796).

Number of observations: 1796

Variable	Mean	Std. Dev.	Min	Max
Price	255,967.1	220,181.7	10,000	3,000,000
Price sqm	2,320.14	882.4023	500	7,571
Services (500m)	0.617	0.486	0	1
Historical building	0.104	0.306	0	1
Distance to city centre (1km)	0.830	0.376	0	1
Green (500m)	0.538	0.499	0	1
Transport stop (500m)	0.992	0.091	0	1
Distance to train station (500m)	0.919	0.273	0	1
Car park	0.058	0.234	0	1
Age*	40.812	24.845	6	219
View of river or mountains	0.765	0.424	0	1
Commercial	0.174	0.379	0	1
Floor level	2.370	1.977	0	9
Walk score	0.499	0.370	0	1
Number of baths	1.722	1.347	1	9
Total surface	104.009	53.595	20	650
Offer sheet (12 months)**	0.723	0.447	0	1
"Movida" 50 m	0.450	0.984	0	6
"Movida" 51-75 m	0.450	0.948	0	7
"Movida" 76-100 m	0.693	1.287	0	9

* Age is defined as the difference between the date on which the residential property was sold and the date of its eventual renovation

** This dummy indicates whether the dwelling has been on offer in the last year

(R^2=0.53) and coefficients have the expected sign. The estimated coefficients are stable across the models and the F-statistic rejects the null hypothesis that all parameters are jointly equal to zero at the 1% level. The analysis of simple correlation matrices indicates that there are no significant dependencies between the variables, and the variance inflation factor (VIF) test confirms that there is no multicollinearity. Finally, the Breusch–Pagan test cannot reject the null hypothesis of constant variance.

Hedonic pricing models usually refer to the noise sensitivity depreciation index (NSDI), which is the percentage change in house prices per dB increase in noise level according the definition of European Union (2002).

As a precise noise phonometric measurement for recreational noise is not available, we proxied nighttime annoyance by the number of open night commercial activities (pubs, night-clubs, and bars) within 100 meters. In particular, we analyzed the effect of noise annoyance within 50, 75, and 100 meters from the dwellings.

Our findings show that the indicator is significant in the range 51-75 meters, but it is not within 50 meters and from 76 onwards. The lack of effect within 50 meters could be due to measurement errors. The coefficient of -0.023 implies a price reduction of -2.7% for each night activity, which means that one additional commercial night activity between 51 and 75 meters bears a price reduction of 2.7%.

Table 3 reports results for log-lin quantile regression model. It shows that in lower quantiles, meaning low price dwellings, the effect of "Movida" noise is not significant, while in higher quantiles it becomes significant and increasingly relevant. In particular, the presence of "Movida" activity within 51-75 meters from the dwelling implies a price reduction of -2.66% for squared meter in the 50th quantile and -3.63% in the 75th quantile.

Table 2.   Regression results for the central subsample

VARIABLES	(2) log-lin
Services	0.376***
	(0.016)
Historical building	0.673***
	(0.029)
Distance to city centre (1km)	0.006
	(0.023)
Green (500m)	0.072***
	(0.014)
School in 500 m	0.035
	(0.031)
Transport stop (500m)	-0.033
	(0.073)
Distance to train station (500m)	-0.071***
	(0.027)
Car park	0.142***
	(0.029)
Commercial	0.196***
	(0.019)
View of river or mountains	0.274***
	(0.025)
Walk score	0.289***
	(0.033)
Number of baths	0.006
	(0.006)
Offer sheet (12 months)	0.147***
	(0.017)
Age	0.000
	(0.000)
Total surface	-0.000
	(0.000)
Floor level	0.005
	(0.004)
"Movida" in 50 m	0.009
	(0.007)
"Movida" in 51-75 m	-0.023**
	(0.010)
"Movida" in 76-100 m	0.004
	(0.005)
Constant	6.905***
	(0.079)
Observations	1,796
R-squared	0.523

*** p<0.01, ** p<0.05, * p<0.1
Standard errors in parenthesis

Table 3.   Effect of "Movida" (51-75 meters) on prices in the central subsample, different specifications.

Model specification	Sign	Coeff	ΔP/P (%)
Model (log-lin)	(-)	-0,023	-2,27%
Model (log-lin) - q50	(-)	-0,027	-2,66%
Model (log-lin) - q75	(-)	-0,037	-3,63%

4 CONCLUDING REMARKS

The results obtained by hedonic modelling show that the adverse environment for an apartment located in a "Movida" district will result in a lower market value as compared to an apartment with similar characteristics, except for recreational noise. This occurs because potential buyers reduce their demand, as they discount present value of the costs of annoyance, loss of tranquility, and possible health effects.

REFERENCES

Agenzia delle Entrate – Central Directorate for Cadastral, Cartographic and Land Registration Services, (2018), The Italian Cadastral System. Agenzia delle Entrate.

Andersson, H.; Jonsson, L.; Ogren, M., (2010), Property prices and exposure to multiple noise sources: Hedonic regression with road and railway noise. Environmental and Resource Economics 45, 73–89.

Blanco, J.C., Flindell, I., (2018), Property prices in urban areas affected by road traffic noise. Applied Acoustics 72, 133–141.

Brandt, S.and Maennig, W., (2011), Road noise exposure and residential property prices: Evidence from Hamburg,Transportation Research Part D: Transport and Environment. Volume 16, Issue 1, 23–30.

Chang, J.S.and Kim, D.J. (2013), Hedonic estimates of rail noise in Seoul. Transport. Res. D Transp. Environ, 19, 1–4.

Cooley D., Barcelos P., (2018), googleway: Accesses Google Maps APIs to Retrieve Data and Plot Maps, URL https://CRAN.R-project.org/package=googleway.

Day, B., Bateman, I., Lake, I., (2007), Beyond implicit prices: recovering theoretically consistent and transferable values for noise avoidance from a hedonic property price model. Environ. Resour. Econ. 37, 211–232.

Del Giudice and others, (2019), The Monetary Valuation of Environmental Externalities through the Analysis of Real Estate Prices, Sustainability, 11(11), 3110.

Eyraud L., (2014), Reforming Capital Taxation in Italy, International Monetary Fund, (2014), WP 14/6.

EUROPEAN UNION (2002), The State-Of-The-Art on Economic Valuation of Noise, Final Report to European Commission DG Environment.

European Commission, 2003. Position Paper on the Valuation of Noise (Brussels).

Hadfield P., (2017), Evening and Night-time Economy, Research, http://www.philhadfield.co.uk/otherpublications.aspx.

Handy, S., (1993). Regional Versus Local Accessibility: Implications for Nonwork Travel, Transportation Research Record, 1400: 58–66.

Knight J.R. (2008) Hedonic Modeling of the Home Selling Process. In: Baranzini A., Ramirez J., Schaerer C., Thalmann P. (eds) Hedonic Methods in Housing Markets. Springer, New York, NY.

Navrud, S., (2002), The State of the Art on Economic Valuation of Noise; Final Report to European Commission DG Environment; Department of Economics and Social Sciences, Agricultural, University of Norway.

Navrud, S. (2004). The economic value of noise within the European union: a review and analysis of studies. Acústica, 14–17.

Nelson J. (2004), Meta-analysis of airport noise and hedonic property values: problems and prospects. Journal of Transport Economic Policy, 38(1):1–28.

Orr AM, Dunse N, Martin D, (2003), Time on the market and commercial property prices. Journal of Property Investment and Finance, 21(6):473–494.

Osservatorio del Mercato Immobiliare, (2019), Statistiche regionali. Il mercato immobiliare residenziale. Piemonte. Agenzia delle Entrate.

Ottoz E., Rizzi L. and Nastasi F., (2018), Recreational noise: Impact and costs for annoyed residents in Milan and Turin, Applied Acoustics, 133. 173–181.

Pivo G.and Fisher J., (2011), The Walkability Premium in Commercial Real Estate Investments, Real Estate Economics 39(2):185–2.

Rahmatian M. and Cockerill L., (2004), Airport noise and residential housing valuation in southern California: a hedonic pricing approach, International Journal of Environmental Science and Technol, 1 (1). pp. 17–25.

RStudio Team, (2015), RStudio: Integrated Development for R. RStudio, Inc., Boston, MA URL http://www.rstudio.com/.

Sirmans S., Macpherson D. and Zietz E., (2005) The Composition of Hedonic Pricing Models. Journal of Real Estate Literature: 2005, Vol. 13, No. 1, pp. 1–44.

Trojanek R. et al., (2017), The impact of aircraft noise on housing prices in Poznan, Sustainability, 9, p. 2088.

WHO Regional Office for Europe (2011), vii. Burden of disease from environmental noise WHO,15.

Soft mobility and perception of urban landscape

Pedestrians, Urban Spaces and Health – Tira, Pezzagno & Richiedei (eds)
© 2020 Taylor & Francis Group, London, ISBN 978-0-367-46171-3

Improving the walkability for next-generation cities and territories, through the reuse of available data and raster analyses

A. Cittadino, G. Garnero & P. Guerreschi
DIST, Università degli Studi di Torino, Italy

E. Eynard & G. Melis
LINKS Foundation, Italy

F. Fiermonte & L. La Riccia
DIST, Politecnico di Torino, Italy

ABSTRACT: The walkability of the city has been the subject of consideration for at least fifteen years, but today it is certainly in evidence for many reasons. Among these, its global relevance for liveable, healthy and resilient cities. This is why it requires a clear operational objective, in order to support the different subjects[1] involved in the city policies and planning. The contribution proposes a two-level reasoning. At the city level, the goal is to recognize the parts where actions aimed at improving walkability can be more effective. The second level is more detailed: lacking pedestrian paths' specific graphs, the urban space is modelled considering a series of criteria and an "impedance" has been assigned to each cell (i.e., the cost of travelling the cell on foot). This approach is applied to the city of Torino (Italy), but it is largely generalizable. The elaborations are seen as an aid to stakeholders to reason on "walkability" and to compare different points of view in an explicit and articulated way.

1 INTRODUCTION

In February 2018 the volume Pedestrians First (ITDP 2018) was inspiring in particular for two reasons: it is conceived in a context and with a vision absolutely global; it has a clear operational objective, to support the different subjects (policy and decision makers, planners and technicians, pro-active subjects at a local level, etc.) involved in the promotion and planning of the increase of the city's walkability, as a factor of sustainability and growth of the its livability.

The city's walkability intersects several relevant current themes. The reasons reported in the literature to deal with it are surprisingly varied, ranging from the sustainability of urban mobility, from soft mobility (recovery of walking as a solution to the problems of transport over short distances), to the health of people (contrast of obesity; Agampatian 2014) prevention of cardiovascular diseases, osteoporosis (Eynard, Melis and Tabasso 2017). Moreover, they can affect air quality and noise pollution and have greater access to green areas, promote neighborhood redevelopment actions, social cohesion and community identity and, finally, widen social networks (Edwards, Tsouros 2006; Cavill et al. 2006). Some analyzed researches often describe a path that led to the construction of walkability indices, taking into consideration, along with other aspects, such as the physical form of the city and its way of functioning, that is population, building density and the mix of urban activities and functions (which, together, lead to multiplication of possible origins and destinations of movements), security

---

1. policy and decision makers, planners and technicians, pro-active subjects at a local level, etc.

(which concerns both the intersections between pedestrian paths and vehicular routes, that anthropic safety), the pleasantness of the environment (quality of the sidewalks, presence of shops and other activities along the pedestrian paths, presence of green, low level of pollution and noise, etc.). If the ultimate goal is to provide views of the walkability of the city, useful for people to reconsider the possibility of walking different areas of the city, and the accessibility of different interest points on foot, the specific objective of this paper is to demonstrate the possibility of constructing effective views of city's walkability, using existing data, extensible (more or less easily) to the entire urban or metropolitan area, bypassing the lack of specific networks related to foot mobility, which allow municipal technicians and policy-makers to focus on critical points of pedestrian paths, considering all the factors that influence the walkability.

2 EVALUATION OF WALKABILITY AT THE CITY LEVEL: WALKABILITO

LINKS developed a WebGIS tool called WalkabiliTO that allows to calculate the urban walkability index. In particular the model has been tested on Turin and is based on pedestrian accessibility to basic services and may be personalised according to customer requirements. This allows to identify which areas offer best conditions to be lived moving on foot, thus promoting a sustainable way of life. Moreover, having the possibility to verify the walkability index for each own home address, the citizen can actually realise which services are effectively reachable on foot without the need for using the car (Figure 1).

The creation of the model for index calculation is based on three steps:

- Data acquisition and optimisation (most data come from the 'Geoportale del Comune di Torino');
- Creation of coverage area (service area) where services are reachable within 400 meters walk;
- Building of the system for weight assignment dependent on the various destinations of interest for the different types of customers.

The data collected was optimized for analysis and grouped into different categories: food, mobility, urban green, loisir and other services. In particular, the following destinations were selected:

Figure 1. Example of instrument query.

- Groceries: convenience stores, supermarkets, food shops, fruit and vegetables, bakeries, butchers, hypermarkets, delicatessens, pastry shops, soft drinks, etc.
- Extra food: hairdressers, bars, hardware, newspapers, tobacco, etc.
- Mobility: local public transport, underground, etc.
- Urban green: gardens and parks.
- Loisir: libraries, cinemas, theaters, sports facilities.
- Other services: primary schools, municipal and affiliated nurseries, public health care offices, hospitals, pharmacies, post offices.

The result is the urban walkability index which can be personalised by assigning different weights to the different destinations of interest that represents for each pixel of the area the degree of pedestrian accessibility visualised in a semaphoric scale on a 400x400 meters cell grid. The tool is active and reachable at the following address: http://www.urbantoolbox.it/pro ject/walkability. A significant feature of the index proposed is represented by its replicability to other Italian cities and, in particular, the possibility of calibrating the instrument with respect to the target group of people involved in the study (young people, the elderly, the disabled, etc.).

3   EVALUATION OF WALKABILITY AT THE NEIGHBORHOOD LEVEL

The approach proposed by the group of the Polytechnic and the University of Turin is applied to the neighborhood level of Turin, but it is broadly generalizable too. On an experimental level, however, we limited the analysis on the area of "San Salvario" district of about 193 Ha. The starting point was the so-called "Walkability Hierarchy of Needs Pyramid" (ITDP 2018), which proposes 6 general and compact criteria in 3 macro-indices: feasibility, safety (physical safety and anthropic safety), comfort/pleasure. The weighted sum of the 3 macro-indices constitutes the Cost Raster (CR), readable as a detailed representation of walkability. High values of the macro-indexes mean high feasibility, high security, high pleasantness, while the CR values represent an "obstacle" to well-walk through the cell (Figure 2).

The values of the CR were then calculated as completion at 100 of the normalized sum of the macro-indices. In addition to the criteria considered, other criteria were also identified that would be reasonable considered, currently not implemented because the necessary data were not found.Another representation of walkability concerns the accessibility to points of interest (POI) within a walkable distance. The Cost Distance (CD) represents the "cumulative cost" of the movement, where the cost is given by the value of the CR cell. This value is multiplied by the size of the cell (in our case, 1 meter) or by its diagonal if the cell is crossed diagonally. So the CD calculates a weighted accessibility that, in our case, takes into account distance, walkability, safety and pleasantness of walking. With the CD method, accessibility was calculated with respect to some of the attractive mobility activities on foot, considered at

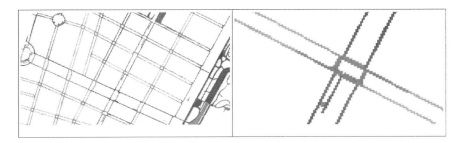

Figure 2.   On the left, the CR: the colour from blue to red, passing through green yellow orange, represents increasing "impedances". On the right, zooming, you can see the 1x1 m raster cells.

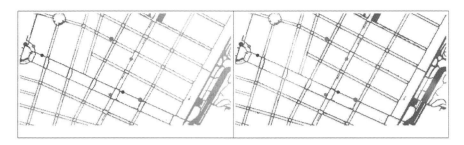

Figure 3. On the left, access to points of intermodality in urban mobility: blue, green, yellow and orange represent decreasing levels of accessibility. On the right, in green the normalized weighted distance is less than the corresponding physical distance, in grey, are more or less equal, in red the weighted distance is greater than the physical distance.

the city level, and for significant points in urban mobility. The weighted sum of several CDs can lead to synthetic views of accessibility.

Figure 3, to the left, represents the weighted sum of the CDs relative to points of intermodality of urban mobility (between [] the weight attributed in the weighted sum): metro stops [50%], bike sharing stations [20%] and car sharing [15%]. An interesting secondary product of the CD algorithm is the so-called *allocation*, i.e. the identification of the area of influence of each activity: the result can be usefully interpreted if it refers to a limited number of destinations.

4 CONCLUSIONS

The described analyses seem to be, and partly proved to be, flexible and effective tools for reasoning on the walkability of the city: the approach is objective, in the sense that the views of walkability are based on data, but we know that they depend to a large extent on implemented indices and weights. Perhaps even more questions and suggestions can arise when looking at the CDs. If the San Salvario test area is limited, the extension of the survey to the whole city, and also to the urbanized parts of the metropolitan area, is feasible, not without costs but with sustainable costs. In fact, we think that the elaborations described in this paper have to be seen as an aid to the various stakeholders to compare different points of view in an explicit and articulated way. Many steps of elaborations are questionable, in the sense that they imply choices, evaluations: it makes sense therefore, precisely, to use these elaborations to discuss on it. Among the desirable developments there is a certain automation of (at least) parts of the procedures, so that they can be used live, in operational contexts or in communication and participation situations. Some automation is also essential to produce sensitivity tests on the various indices and weights used, tests that help to focus on what is really important.

REFERENCES

Agampatian R., (2014). *Using GIS to measure walkability: A Case study in New York City. Master's of Science Thesis in Geoinformatics, Royal Institute of Technology (KTH) Stockholm, Sweden. Retrieved from* https://pdfs.semanticscholar.org/20ef/6302095148094b97b229c50fe236321be338.pdf. *Last accessed May 2019.*

Cavill, N., Kahlmeier, S., & Racioppi, F. (Eds.). (2006). Physical activity and health in Europe: evidence for action. World Health Organization.

Edwards, P., & Tsouros, A. D. (2006). Promoting physical activity and active living in urban environments: the role of local governments. WHO Regional Office Europe.

Eynard E., Melis G. & Tabasso M., (2017). Una metodologia per il calcolo della camminabilità in città: il walk index a Torino. In: ASITA 2017. 477–482.

ITDP, Institute for Transportation and Development Policy, (2018). *Pedestrians First, Tools For a Walkable City*, *ITDP, New York*, www.itdp.org.

La Riccia L.,Cittadino A., Fiermonte F.,Garnero G., Guerreschi P., Vico F. (2019). The Walkability of the Cities: Improving It Through the Reuse of Available Data and Raster Analyses. In: Voghera A., La Riccia L. (eds.), Spatial Planning in the Big Data Revolution, IGI Global, Hershey PA, USA. 113–137.

Pedestrians, Urban Spaces and Health – Tira, Pezzagno & Richiedei (eds)
© 2020 Taylor & Francis Group, London, ISBN 978-0-367-46171-3

Investigating the importance of walk stages as a factor in the choice between car and public transport in urban areas

D. van Soest, M.R. Tight & C.D.F. Rogers
University of Birmingham, UK

ABSTRACT: The distances people walk to and from public transport tend to differ a lot across urban areas. This study tries to shed more light on this variability by collecting detailed smartphone tracking data of participants from two areas in Birmingham, UK. The areas are distinctive for their public transport provision. Besides walks related to public transport, the study also considered walks for potential public transport trips if all car trips would be replaced. There were no significant differences between the areas in the public transport-related walk distances. However, when considering the walks for potential public transport trips, people in the area without railway access would require longer walks. Also transport attitudes of people seemed to be related to the potentially required walk distance.

1 INTRODUCTION

One way to decrease many of the adverse impacts of private car use in urban areas is to promote a shift towards public transport (PT). This can also enhance health and well-being, because it usually requires access transport, which, certainly in urban contexts, is frequently done by foot or bicycle. When access and egress trip stages, i.e. the trips to and from PT, become too long, travellers are likely to choose other transport modes (Krygsman, Dijst, & Arentze, 2004). The distances people walk to access PT can vary significantly across areas and depend on a wide range of factors (van Soest, Tight, & Rogers, 2019). Although in transport planning frequently adopted guidelines suggest that people are willing to walk 400m for buses and 800m for rail transport (e.g., Canepa, 2007), in some cases walk distances were found to be up to almost 1400m on average (Jiang, Zegras, & Mehndiratta, 2012).

These differences are partly caused by factors that determine the spatial coverage of PT, such as the routes, stop spacing, land use and density. However, the attractiveness of both the PT system and the alternative travel options available, in particular the car, plays an important role as well. For instance, travellers are willing to walk further to trains than buses (e.g., Alshalalfah & Shalaby, 2007) and to more frequent services (e.g., Mulley, Ho, Ho, Hensher, & Rose, 2018).

Although the level of PT services available to travellers, especially the density of routes and stops, seems to have an important influence on the distances people walk, it is rarely accounted for explicitly in studies, which tend to focus on people at certain stations or on PT services. Moreover, the focus on walk distances of current PT users lacks insights on walking acceptability from those who do not yet use PT, whilst the ultimate goal is often to establish a modal shift from private car to PT. Finally, most existing research on walk distances to and from PT has used self-reported methods, whilst more objective and detailed tracking technologies are available nowadays. Therefore, the goal of this paper is to study the variation in walk distances to and from PT using detailed objective data and explore how the acceptability of walking and PT as a transport mode differs between urban areas with different PT provision and for people with different attitudes towards transport.

2 RESEARCH APPROACH

The research approach involved the collection of both subjective and objective data related to travel behaviour. Firstly, a postal questionnaire survey was conducted to obtain insights into perceptions of transport and the barriers experienced when combining walking and PT as a means of travel. In the second phase, objective travel behaviour data were collected by tracking trips using a data collection app for smartphones. These participants were gathered through the questionnaire, in which they could state their interest in the follow-up study.

The questionnaire was distributed to households in four areas of Birmingham. The areas were mainly distinctive on their level of deprivation and access to railway transport. From the 4000 questionnaires distributed (1000 in each area), 500 valid responses were received. 45 participants, all living in the two less deprived areas, additionally tracked their trips. The apps and back-end system used for the tracking study were based on the open source system MEILI (Prelipcean, Gidófalvi, & Susilo, 2018). The location points collected were annotated manually, resulting in 827 trip legs.

3 RESULT OF THE RESEACH

The respondents to the questionnaire were clustered based on their attitudes towards walking, cycling, car driving, bus use and train use. For each mode they rated a simple question to what extent they liked using the mode on a 5-point Likert-type scale (e.g. "I like walking"). A hierarchical clustering method was applied, after which four attitudinal clusters were obtained. Cluster 1 had a positive attitude to all modes, except buses. Cluster 2 is relatively positive about buses, but less positive about car driving. Cluster 3 was negative about cycling and bus use. Cluster 4 is less positive to walking and PT compared to the other clusters.

Of the 45 participants in the tracking study, there were 15 people who recorded at least one trip by PT. None of these were in cluster 4. For most of the PT trips, a walk to or from the PT mode was registered as well, 81 in total (excluding transfer walks, such as between a bus stop and a railway station). Some of these walks are not independent, though, with similar walks both ways, or repetitive over multiple days. Hence, in the analysis only unique trips are taken into account (e.g. if someone walked from home to the home railway station and vice versa on two days, thus having walked four access/egress trips, only one of these trip legs is counted).

Figure 1 shows the cumulative distribution of walks per distance to/from buses and trains. For trains, the conventional 800m threshold appears to be quite accurate and cover most of the walks registered. About 15% are significantly longer. In contrast, for buses there is actually a large proportion of walks longer than 400m (around 40%), up to almost four times as long. The difference in mean distance between walks to and from buses and trains is not significant in the sample ($t = -1.6112$, $df = 40$, $p = 0.115$). This might be partly due to the small sample size, but also reflects the finding that the walk distances related to bus trips are relatively long. Also between the two areas, differences are not statistically significant (bus trips: $t = -0.17211$, $df = 16$, $p = 0.866$; train trips: $t = -1.0993$, $df = 14$, $p = 0.290$).

A closer look at the tracked routes revealed that the large majority of walks were undertaken via the most direct and shortest route. In case of train journeys, also the closest station was used in most cases. There were a few instances in which a participant chose a station that required a slightly longer walk, but this was in the direction of travel, leading to a shorter overall travel distance. In the case of walks related to buses, the closest bus stop was not necessarily used, but very much dependent on which bus lines operate where.

There were 246 car trips with complete information about their origin and destination. For each of these car trips, it was examined how likely the trip would have been if the person had chosen to use PT instead. This was done using an R-script that could connect to the API (application programming interface) of TransportAPI, to retrieve a travel plan for a trip between the coordinates of the origin and the destination of the car trip. Some car trips also included walk access or egress trips (e.g. from a parking garage to the final destination), in this case the origin and destination of the complete trip were used. To reduce potential

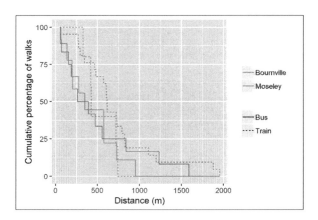

Figure 1. Cumulative distribution of walk distances to/from bus and train.

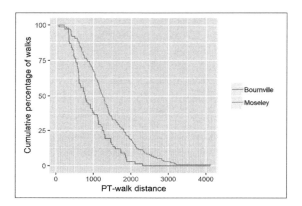

Figure 2. Cumulative distribution for PT-walk distances in alternative PT trips, for both neighbourhoods.

dependency between trips, for trips that were made repeatedly between the same origin and destination by a particular user, only the first trip is used for analysis.

Figure 2 displays the cumulative distance distributions for potential access and egress walks if the car trips had been undertaken by PT, for the two areas. There is a significant difference, with Moseley residents requiring longer walks if they would use PT more often (t = 5.0119, df = 178, p = < 0.001). About 70% of the car trips by Moseley residents would require more than 1000m of walking if made by PT, whilst for the participants in Bournville this is approximately 40%.

When comparing for attitudinal clusters, there were significant differences between clusters 4 and 1 (t = -2.14, df = 28, p = 0.041) and between clusters 4 and 3 (t = 2.2127, df = 30, p = 0.035); cluster 4 participants require significantly larger walk distances if they were to shift to public transport.

4 CONCLUSIONS

The objective of the research was to study the variation in walk distances to and from PT using detailed objective data and explore how the acceptability of walking and PT as a transport mode varies between urban areas with different PT provision and for people with different attitudes towards transport.

Using a smartphone tracking study, the trips of 45 participants were recorded in an objective and detailed way. It became clear that the walk distances to and from trains were quite well in accordance with the conventional threshold of 800m. The walk distances for trains did not differ significantly with walks for buses, however; many bus walks were much longer than the conventionally assumed 400m. This might be due to the layout of bus routes in Birmingham in relation to where people live, meaning the desired bus route is often not served by the closest bus stop. Whilst this can be a barrier for uptake of PT, it does enhance physical activity of the existing PT users.

Examination of car trips and their PT alternative revealed that a large share of the trips would require more than 1000m of walking (access + egress) if made by PT, especially in Moseley. In Moseley, the share of PT trips was already lower, so it indicates a lack of access to PT for the trips required.

In the next phase of the project, more detailed relationships with the characteristics of PT provision will be explored. This should subsequently give better insights in the role of the required access walks in the choice for PT instead of the car.

REFERENCES

Alshalalfah, B. W., & Shalaby, A. S. (2007). Case Study: Relationship of Walk Access Distance to Transit with Service, Travel, and Personal Characteristics. Journal of Urban Planning and Development, 133(2),114–118.

Canepa, B. (2007). Bursting the Bubble. Transportation Research Record: Journal of the Transportation Research Board, 1992(1),28–34.

Jiang, Y., Zegras, P. C., & Mehndiratta, S. (2012). Walk the line: station context, corridor type and bus rapid transit walk access in Jinan, China. Journal of Transport Geography, 20(1),1–14.

Krygsman, S., Dijst, M., & Arentze, T. (2004). Multimodal public transport: an analysis of travel time elements and the interconnectivity ratio. Transport Policy, 11(3),265–275.

Mulley, C., Ho, C., Ho, L., Hensher, D., & Rose, J. (2018). Will bus travellers walk further for a more frequent service? An international study using a stated preference approach. Transport Policy, 69(January), 88–97.

Prelipcean, A. C., Gidófalvi, G., & Susilo, Y. O. (2018). MEILI: A travel diary collection, annotation and automation system. Computers, Environment and Urban Systems, 70(December 2017), 24–34.

van Soest, D., Tight, M. R., & Rogers, C. D. F. (2019). Exploring the distances people walk to access public transport. Transport Reviews.

Pedestrians, Urban Spaces and Health – Tira, Pezzagno & Richiedei (eds)
© 2020 Taylor & Francis Group, London, ISBN 978-0-367-46171-3

Moving through the Quarries Park. The case of Brescia

M. Tononi & A. Pietta
Department of Economics and Management, University of Brescia, Italy
IRIS - Interdisciplinary Research Institute on Sustainability, Brescia, Italy

ABSTRACT: Since the Eighties of the last century European industrial cities have lived deep transformations which changed their economic model. Such transformations, and the environmental problems left on the ground, imply the need to rethink the urban model towards higher levels of sustainability. Until some years ago the peri-urban spaces in the southeastern area of Brescia were characterised by the exploitation of gravel mining. Today, through a participatory process driven by the population these spaces are becoming the biggest urban park in the city. The methodology used to study this landscape transformation is based on a participatory approach, considering in particular the participatory action research - PAR (Kindon et al., 2007). We created a participatory map of the cultural ecosystem services in the Park looking at the perception of the stakeholders, focusing on the local community. The analysis also includes data about the accessibility to these cultural ecosystem services.

1 INTRODUCTION

The Province of Brescia is one of the most developed industrial regions in the Northern Italy. The image of the city has been influenced by the presence of the metalworking factories within its borders and in the three valleys in the north (Trompia Valley, Sabbia Valley, Camonica Valley). The history of metalworking in the area started in the pre-roman period and passed through the Roman emperor, Lombard times and the Venetian domination. During the Sixties of the last century the industrial expansion reached its maximum historical level, with over 40.000 employees in the manufacturing process only in the city area, where metalworking represented 61% of the whole industrial sector (Tallone, 1976). In the last decades the relevance of this sector has decreased due to a number of factors like the gradual tertiarization of the city's economy, delocalization processes and the global economic crisis. Today the industrial past can be recognised looking at disused factories, abandoned industrial areas as well as at brownfields sites among them gravel quarries. In some cases former industrial areas became residential ones while a few areas are now involved in re-naturalization processes. This is the case of the subject of this paper, the "Parco delle Cave" (Quarries Park).

2 OBJECTIVE OF THE PAPER

The Quarries Park is an urban re-naturalized area. The local community fought for many years to obtain the institution of a park and avoid new exploitation forms of the area like landfills. The area is very important from an ecological point of view because of the ecosystem services it provides, like food, water and other products provisioning, temperature regulating (so contrasting the urban heat island effect), air cleaning, biodiversity enhancing and a number of cultural ecosystem services. This paper focuses on the cultural ecosystem services (CES) (Millennium Ecosystem Assessment, 2005) inside the park. This analysis developed by the geography research group at the University of Brescia is part of the interdisciplinary project "Parco delle Cave. Un cuore blu in città" funded by Cariplo Foundation. The project is led by the local social cooperative Cauto and involves a number of partners: institutional partners like the Local Municipality, the University, many schools, and the local community together with environmental and sports associations.

Our analysis has a couple of goals. The first is to reconstruct the evolution of the landscape in the area mapping the cultural ecosystem services according to the viewpoint of the local community. The second goal is to analyse the accessibility to these mapped places considering the ways to arrive into the park (by car, public transport, bike or on foot), and the ways to move inside the park area. We paid particular attention to the accessibility of children, elders and people with reduced mobility.

3 METHODOLOGICAL APPROACH AND RESEARCH APPROACH

We developed this study starting from the framework proposed by Fish, Church and Winter (2016) on the identification of places in which a community doing some practices can receive some benefits, mainly intangible, provided by CES. We are investigating CES using a participatory mapping process (McLain et. al., 2013): participants indicate the places inside the Quarries Park in which some practices give them benefits in terms of cultural experience.

The Ecosystem Services are the benefits obtained from the ecosystems and are divided in supporting, provisioning, regulating and cultural services (Millennium Ecosystem Assessment, 2005). Services of the first three categories provide material benefits quantifiable using quantitative approaches. On the other hand, cultural ecosystem services are non material benefits as "cultural diversity, spiritual and religious values, knowledge systems, educational values, inspiration, aesthetic values, social relations, sense of place, cultural heritage values, recreation and ecotourism" (Millennium Ecosystem Assessment, 2005). So, to measure CES it is necessary to adopt different approaches, also involving citizens. For this reason the research adopts a participatory approach for mapping the local community perceptions of the CES in the Quarries Park. We decided to use the Participatory action research methodology (Kindon et al., 2007) to involve the community in some working groups. The participatory action research (PAR) looks both at the quality of information generated and at the ways in which skills, knowledge and participants' capacities are developed through the research experience (McLain et. al., 2013). So, it allows us to reflect with the local community on the evolution of the landscape in the Quarries Park from a personal and systemic-relational perspective considering a range of alternative ways of thinking and acting to valorise the area. This is the second project we promote in the area using this approach. The first one revealed a community ready to collaborate on sustainability topics (Tononi et al., 2017).

4 DESCRIPTION

The first step of the analysis consists in some interviews to the local historians about the evolution of the landscape and the beginning of gravel mining. This is important to have a picture of the evolution of this territory in the last seventy years. The second step is represented by the involvement of local associations and groups in some laboratories of CES mapping to coproduce a first map about the places to which people recognize cultural values. This is the map that we present in this paper. The third step, that we will develop in the coming months consists in the involvement of the community in some public labs of CES mapping to complete the map and discuss alternative governance methods of the park. The Local Neighborhood Councils are helping us to involve the population.

4.1 *Innovative aspects*

The participatory mapping allows us to promote participation, acquire information (Craig et al., 2002; Elwood, 2010; Burini, 2007; Casti, 2007; Brown, 2012; Brown, Kytta, 2014;

Capineri et al., 2016; Goodchild, 2007) and translate our results on CES in a digital map to share with stakeholders, citizens and experts. Producing a participatory map means not only to draw and shape the results on it but also build a process that is part of this map. Today, with participatory methods and technologies, the map is conceived as a process with different steps and participants involved. This process promotes a mutual exchange of information, opinion and knowledge among participants. For experts this means dealing with the local knowledge of the landscape for a better planning; for the local community it means to promote awareness of the cultural elements of the local landscape.

5 RESULTS OF THE RESEARCH

To collect data we prepared a form in which participants have first to indicate the places on the park's map and then to answer to a number of questions describing their interaction with these places and the cultural services they provide (e.g. in terms of identity, experiences, things they can do, etc.). There are also some questions about mobility and accessibility looking at how to reach the park and the selected places.

At present we are developing the second step of the project and so far we have mapped 21 places with a cultural value recognised by the community (Figure 1). Looking at the accessibility, together with the Consulta per l'Ambiente (Council for the Environment[1]) we have produced a map (Figure 2) focusing on the roads (the red lines indicate roads open to all the vehicles) and the bike lanes (the purple and the blu lines indicate paths open only to bikers and pedestrians) inside the park. This map can also be used by citizens and tourists to visit the Quarries Park by bike or on foot.

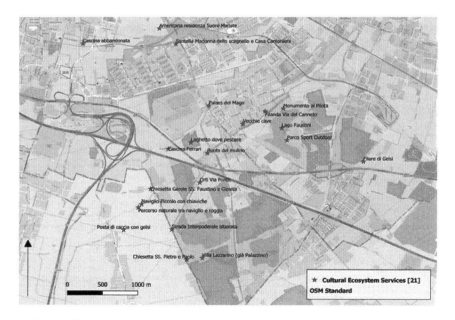

Figure 1.   Places indicated during the participatory mapping of cultural ecosystem services.

---

1. The Consulta per l'Ambiente was set up as an umbrella organisation to bring together associations linked to the environment and its protection, with well defined rules of enrolment, participation and operation. It was created by Brescia Town Council.

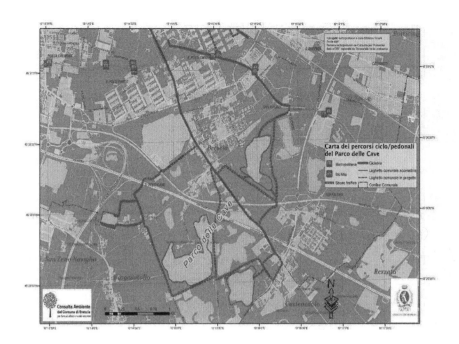

Figure 2. Cycle paths inside the Quarries Park.

Linking the data of these maps it emerges that the accessibility to and inside the park is very complex to manage. Until a few years ago the area was characterised by the exploitation of gravel mining and the Park was formally instituted in 2016. Moreover, the property is mixed public and private and the bike paths don't cover the entire perimeter. Accordingly, people reach the park (and in some cases also the mapped places) mainly by car and then go around inside on foot. Only a few people use exclusively public transport, bike or go on foot. Now the Local Administration is acquiring some areas around the lakes and improving the accessibility reducing the slope and differences in height and opening some lanes only to pedestrians and bikers. In addition, a number of local associations and cooperatives are developing projects with people with reduced mobility like excursions to fishing ponds and canoeing courses for people with disabilities, visits to the park for kids, didactic activities for the primary and secondary school classes.

6 CONCLUSIONS

First data show the complexity of the area looking both at the different types of nature shaping the mapped places and the ways to reach them. The preliminary data show that the local community recognises the presence of CES in different types of landscapes that we have categorised into:

- Agricultural landscape: crop, stream, trees along the streams, "cascine" - old typical buildings of the lombard agricultural landscape;
- Urban Nature: green paths and gardens of ancient historical buildings;
- Leisure natural landscape: the re-generated places around the quarry lakes;
- Industrial landscape: signs of the excavation or other industrial activities like a spinning mill, a water mill, with a value in terms of industrial archeology.

These different elements are part of the natural and human processes that during the years form the complex landscape of the "Parco delle Cave". The data collection will continue in the next few months and we will elaborate a broader categorization of the CES of the area.

Looking at the accessibility the situation is very complex to manage due to a number of factors and, as a consequence, many CES indicated by the local community are not accessible by a part of the population. The initiatives to make it more accessible are only at an experimental level with a few actions led by the Local Administration and the local associations and cooperatives.

We hope that future planning will profitably take into account these mapped places with the presence of CES promoting accessible paths to and inside the Park.

REFERENCES

Brown G., (2012). Public Participation GIS (PPGIS) for Regional and Environmental Planning: Reflections on a Decade of Empirical Research. URISA Journal, 25, 2. 7–18.

Brown G., Kyttä M., (2014). Key issues and research priorities for public participation GIS (PPGIS): A synthesis based on empirical research. Applied Geography, 46. 122–136.

Burini F., (2007). Sistemi cartografici partecipativi e governance: dalla carta partecipativa ai PPGIS. In: Casti E., a cura di Cartografia e progettazione territoriale. Dalle carte coloniali alle carte di piano. Torino. Utet.

Capineri C., Haklay M., Huang H., Antoniou V., Kettunen J., Ostermann F. and Purves R., a cura di (2016). European Handbook of Crowdsourced Geographic Information. London. Ubiquity Press.

Casti E., a cura di (2007). Cartografia e progettazione territoriale. Dalle carte coloniali alle carte di piano. Torino. Utet.

Craig W.J., Harris T.M., Weiner D., a cura di (2002). Community Participation and geographic information system. London. Taylor & Francis.

Elwood S., (2010). Geographic information science: Emerging research on the societal implications of the GeoWeb. Progress in Human Geography. 3. 349–357.

Fish R., Church A., Winter M., (2016). Conceptualising cultural ecosystem services: A novel framework for research and critical engagement. Ecosystem Services, 21. 208–217.

Goodchild M., (2007). Citizens as sensors: the world of volunteered geography. GeoJournal, 69, 4. 211–221.

Kindon S. L., Pain R., Kesby M. (2007). Participatory action research approaches and methods. Connecting people, participation and place. Abingdon. Routledge.

McLain R., Poe M., Biedenweg K., Cerveny L., Besser D., Blahna D., (2013). Making Sense of Human Ecology Mapping: An Overview of Approaches to Integrating Socio-Spatial Data into Environmental Planning. Human Ecology, 41. 651–665.

Millennium Ecosystem Assessment, (2005). Ecosystems and Human Well-being: Biodiversity Synthesis. World Resources Institute,.Washington DC.MA.

Tallone O. (1976). Brescia città industriale. Pisa. Giardini editori e Stampatori.

Tononi M., Pietta A., Bonati S., (2017). Alternative spaces of urban sustainability: results of a first integrative approach in the Italian city of Brescia. The Geographical Journal, 183, 2. 187–200.

Pedestrian road safety

Pedestrians, Urban Spaces and Health – Tira, Pezzagno & Richiedei (eds)
© *2020 Taylor & Francis Group, London, ISBN 978-0-367-46171-3*

Methodology for data processing for road accidents that involve vulnerable road users. the case of Brescia

G. Maternini & M. Bonera
Department of Civil, Environmental, Architectural Engineering and Mathematics, University of Brescia, Italy

M.G. Speranza, C. Archetti & M. Martinello
Department of Economics and Management, University of Brescia, Italy

ABSTRACT: Road safety is a major concern that Authorities aim at tackling especially in urban areas, where most of the accidents involving vulnerable road users occur. The availability a fact-finding background and performant tools to support the interventions can be beneficial, also to manage resources more efficiently. This research aims at devising an automatic tool to simplify the identification of critical aspects and associate them with specific interventions to improve road safety. The earlier step consists in collecting and analysing road accident data and then identifying prototypical accident scenarios, based on infrastructure characteristics. Thanks to the latest Agreement signed by the University of Brescia and the Prefecture of Brescia, researchers will have access to the Police records of road accidents occurred in Brescia, to collect data to perform targeted analyses. The paper shows the first results of the research, with a focus on road accidents involving pedestrians and cyclists in the city of Brescia.

1 INTRODUCTION

Road safety is a major concern that Authorities aim at tackling, by promoting actions and policies that are necessary to reduce the number of road accidents, as expected by the EU goal for 2020, namely halving the number of fatalities compared to the number of 2010 (ETSC, 2018). In addition to saving human life, many are the benefits that derive from improving road safety, such as enhancing the attractiveness of the urban environment, the overall quality of life and fostering more sustainable mobility habits, which are generally among the main objectives of the public administrations (Eltis, 2019; Tira, 2018). Also, in terms of resources allocation, improving road safety would lead to strong economic savings: in 2018, Italy had an expense of 18,6 billion euros for road accidents, that corresponds to 1% of the national GDP (Istat, 2019). According to the latest Italian statistics, 126,701 collisions occurred on urban roads in 2018 (74% of the total), which provoked 169,573 injuries and 1,402 fatalities. Vulnerable road users (VRUs)[1] accounted for a 49% of the victims and, compared to other road user categories, starting from 2010 they registered a slighter decrease or even an increase in road accidents. The Province of Brescia, in 2017, held the second place after Milan in terms of road fatalities and injuries in Lombardy, with 3,336 collisions with 63 deaths and 4,763 injuries in total (ISTAT, 2018). Considering the Municipality of Brescia in 2017, 836 road accidents have been recorded with 6 deaths and 1,118 injuries[2]. The complexity of the urban context and the difficulties related to the conflicting needs make it difficult to implement

---

1. Vulnerable road users (non-motorized users) account for pedestrians and cyclists.
2. Road accident per Municipality. ISTAT database 2017: http://dati.istat.it

interventions in such a context. Therefore, for the public Administration, it is extremely useful to have a clear overview of the phenomenon and to obtain helpful tools to support interventions and manage resources more effectively.

2   METHODOLOGY FOR PROCESSING ROAD ACCIDENT DATA

Following the past research activity of the renewed Research Centre CeSCAM[3] of the University of Brescia (Maternini, 1994), researchers are leading a study related to road safety for VRUs. The ultimate aim of the study is to provide local authorities and technicians with an efficient tool to support their activities to improve the safety conditions of roadways design and surrounding land-use. This research aims to apply the theory of *"prototypical accident scenarios"* by employing statistical classification methods, which are based on a *feature selection* algorithm able to automatically identify classes of observations and detect groups of cases with similar features (Bolón-Canedo, Sánchez-Maroño, & Alonso-Betanzos, 2013). A prototypical scenario is defined as "a model of an accident process which is characterised by a chain of fact, actions, casual relations and consequences in term of damage to people" (Fleury, 2012). This inductive approach, which is based on comparisons between cases, allows to identify similarities between several observations to build a "reference collision setting". By knowing the overall structure of an accident scenario, it is possible to recognize situations or features that can favour a road collision. The original theory refers to the accident dynamic[4] while in this research the target is to find scenarios based on features which are attributable to the infrastructure and surrouding land-use. The research is structured in five phases as follows:

1. Data collection and processing;
2. Fact-finding knowledge;
3. Identify prototypical accident scenarios;
4. Develop an automatic algorithm;
5. Devise and associate design solutions.

Since the theory of prototypical scenarios still refers to an a-posteriori approach, it requires historical data to be performed, whereas the proposed procedure in this research will allow a preventive road safety analysis. Although it is possible to freely consult the ISTAT platform, data available are already disaggregated and refer to just some of the information collected at the national level, so that the information provided is lacking and it is not possible to retrace all the single road accidents and their details. The latest Agreement signed by the University and the Prefecture of Brescia[5] (which is one of a kind in Italy so far) was fundamental to carry out the initial phases of the research, in order to collect all the information needed directly from the Police Authorities that are in charge of recording road accidents and create a fact-finding knowledge, necessary to pinpoint the concerns of the case study area. Data will be transmitted to the University through a specific form, designed by the researchers based on the ISTAT template[6]. This paper aims at describing the results of these initial phases.

---

3. Research Centre *"Centro Studi Città Amica per la sicurezza nella Mobilità"* (CeSCAM) of the Department of Civil, Environmental, Architectural Engineering and of Mathematics of the University of Brescia.
4. Accident dynamic is divided into several phases: situation before driving, driving situation, accident situation or kinematic condition, emergency and collision situation.
5. *"Agreement for the prevention of road accidents"*.
6. ISTAT template CTT/INC, version 2018.

3 FIRST STEPS RESULTS: A FACT-FINDING KNOWLEDGE

Identifyng the main features of road accidents, especially the ones related to the infrastructure design, is the expected outcome of the initial phase of the research. To perform the data analysis, the statistical software SPSS and the GIS software QGIS have been employed. A total of 58 events have been recorded in the Municipality of Brescia over the period 2014-2018[7], of which 42 pedestrians (ped.) (10 fatalities and 32 seriously injured) and 16 cyclists (cyc.) (5 fatalities and 11 seriously injured). Figure1 shows the spatial distribution of the accident. It is possible to identify the areas with a higher percentage of collisions, which correspond also to the roads that are the main access to the city centre (Via Triumplina, Via Milano, Viale Venezia and Via Corsica). Tab.I reports in percentage of frequency the main infrastructural features of the sites where accidents occurred, split in pedestrians and cyclists.

As known, most of the collisions occurred along class E roads (68% ped. and 55% cyc.), where the speed limit is 50km/h and traffic is usually higher. Also, two-way roads and the ones with one lane per direction showed a lower level of safety, especially related to the risk when crossing (almost 80% ped. and 62.5% cyc). Most of the accidents occurred along straight segments (71% ped. and 50% cyc.), where usually driver awareness is lower than elsewhere due to the road profile. In these cases the road design would play a core role. Most of the collisions occurred with regulamented signage (85.3% ped. and 68.8% cyc.), so that the current signage method would seem obsolete. It is worthy to report that most of the violations registered were attributable to a wrong behaviour at crossing by pedestrians (19%), the missed right-of-way to VRUs by drivers (27,6%) and wrong driving behaviour and speeding of drivers (20.6%). These observations are meaningful since they highlight that could exist inadequacy of roadway design and conditions, which compromise the readability of roads and therefore their safety conditions.

Fig. 1.   Road accident per type and quarter occurred in the Municipality of Brescia over the period 2014-2018. Data processing with QGIS software.

---

7. Data from the Police of the Municipality of Brescia for fatalities and serious injuries.

Table 1. Percentages of responses related to infrastructure features of the road accident occurred in the Municipality of Brescia (2014-2018).

Variable	Categories	Pedestrian (ped) [%]	Cyclist (cyc) [%]
Road funcional classification[8]	Inter-district urban road (E)	39.0	43.8
	District urban road (E)	29.3	12.5
	Inter-zone urban road (F)	19.5	12.5
	Local road (F)	7.3	18.9
	Other non-urban road	4.9	12.5
Carriageway	Undivided	81.0	75.0
	Divided	19.0	25.0
Road direction	Two-way	76.2	62.5
	One-way	23.8	37.5
Lane per direction	Single lane	81.0	62.5
	Double lane	19.0	37.5
Roadway site	Straight segment	71.4	50.0
	Signalised intersection	14.3	12.5
	Traffic light intersection	9.5	18.8
	Roundabout	2.4	6.3
	Turn	2.4	6.3
	Parking	-	6.3
Crossing nearby	Yes	61.9	37.5
	No	37.1	62.5
Road signage	Vertical	2.4	-
	Horizontal	7.1	18.8
	Horizontal and vertical	83.3	68.8
	None	4.8	6.3
	Temporary works	2.4	-
Obstacles for driver	None	81.0	87.5
	Urban green	2.4	6.3
	Vehicles	14.3	6.3
	Dumpsters	2.4	-
Obstacles for victims	None	95.2	87.5
	Urban green	2.4	6.3
	Dumpsters	2.4	6.3
Traffic condition	Low	23.8	25.0
	Normal	71.4	68.8
	Congestion	4.8	6.3
Visibility	Good	66.7	81.3
	Sufficient	26.2	18.8
	Insufficient	7.1	-

4 CONCLUSIONS

This paper shows the first outcomes of the research conducted in Brescia about road safety for vulnerable road users, to evaluate concerns attributable to road infrastructure and design features. In this phase of the study, researchers collected records just related to fatalities and seriously injured, therefore it would be better to have access to more road accident data (e.g. all the injuries), in order to apply the Cluster Analysis. Beside this, the results obtained high-light the strong role that infrastructure plays on road collision occurrence.

Future research will be devoted to developing a technology capable of identifying the most prominent attributes associated with collisions occurrence through the application of feature

---

8. Functional classification provided by the Municipality of Brescia in the Mobility System Plan of the Local Development Plan (*PGT Brescia – "Assetto Mobilità. Classificazione funzionale Stato di Fatto" Attachment ALall01b*).

selection techniques. Successively, classification algorithms will be applied on these attributes in order to classify the accidents on the basis of the users involved and their seriousness.

REFERENCES

Bolón-Canedo, V., Sánchez-Maroño, N., & Alonso-Betanzos, A. (2013). A review of feature selection methods on synthetic data. Knowledge and Information Systems. doi:10.1007/s10115-012-0487-8.

Eltis. (2019). Guidelines for Developing and Implementing a Sustainable Urban Mobility Plan (Second Edition).

ETSC. (2018). 12th Road Safety Performance Index Report. Bruxelles.

Fleury, D. (2012). Sicurezza e urbanistica. L'integrazione della sicurezza stradale nel governo urbano (Gangemi Ed). Rome.

Istat. (2019). Incidenti stradali in Italia - Anno 2018.

ISTAT. (2018). L' incidentalità sulle strade della Lombardia nel 2017.

Maternini, G. (1994). La sicurezza del pedone in città. Il caso di Brescia (Vol.1-Al). Brescia: Sintesi Editore.

Tira, M. (2018). A safer mobility for a better town: the need of new concepts to promote walking and cycling. In M. Tira & M. Pezzagno (Eds.), Town and Infrastructure Planning for Safety and Urban Quality: Proceedings of the XXIII International Conference on Living and Walking in Cities (LWC 2017), June 15–16,2017, Brescia, Italy (p. 442). Brescia: CRC Press.

Bradley, P.S., Mangasarian, O.L., & Street, W.N (1998). Feature selection via mathematical programming. INFORMS Journal on Computing, 10:209–217.

Bertsimas, D., & Dunn, J. (2017) Optimal classification trees. Machine Learning, 106: 1039–1082.

Analyses of factors influencing children behaviour while crossing the conflict zones at urban intersections

I. Ištoka Otković
Faculty of Civil Engineering and Architecture, Josip Juraj Strossmayer University of Osijek, Croatia

A. Deluka-Tibljaš & S. Šurdonja
Faculty of Civil Engineering, University of Rijeka, Croatia

A. Canale, G. Tesoriere & T. Campisi
Faculty of Engineering and Architecture, University of Enna KORE, Italy

ABSTRACT: Statistical analyses at World level show that traffic accidents are main reasons of death for adolescents and second reason for death of children aged 5 to 14 years. Pedestrian crossings are areas where there is high risk of conflict between pedestrians and vehicles and this is why in this paper research and analyses of children behavior while crossing intersection conflict area (street) is done. Field data were collected in three different cities: Osijek and Rijeka in Croatia and Enna in Sicily, Italy. The aim was to establish what and how influences children behavior while crossing the street at a signalised intersection. Different parameters regarding both children and their behavior and street conditions were registred for approximetly 300 children in every city and correlated to the time spent while crossing the street.

1 INTRODUCTION

Traffic accidents are main reason of death among adolescents and second reason for children aged 5 to 14 years. Analyses done at World level show that children are injured as pedestrians at higher rates than other transport modes: about 38% of child mortality in road crashes is as pedestrians (WHO, 2015). Reasons and potential measures to mitigate this sever traffic safety problem are undertaken for decades. Complex study done on children's perception of safety and danger on the road (Ampofo-Boateng et all, 1991) showed that children aged 5-7 are not able to judge well potentially dangeorus traffic situations while those aged 9-11 demonstrate fare more ability to detect traffic danger. In the same time no sex differencies were apparent. Some recent studies concerning gender (Wang et all, 2018.) and age influence (Wang et all, 2018; Chang et all, 2018) on children's pedestrain behaviour proved that older children behave more safely than younger ones and that girls tend to have safer behavior with increasing age in greater extent than boys. It was also established that different are the risky behavior depending on the age – younger children do have problem in perception of traffic hazards but olders were more likely to violate a basic pedestrian behaviour rules and in some cases their behavior can be pointed as more risky (Gitelman et all, 2019). Concerning different potentially risky behavior among boys and girls it was established (Wang et all, 2018) that girls prefer to walk with partner what can lead to neglect traffic risks and boy's risks are more connected to their active, impulsive and playful behaviors. Although similar, from selected researches can be concluded that children's traffic behavior does not have the same patern in different countries (Gitelman et all, 2019., Wang et all, 2018.). In order to define most effective mitigation measure for every place it is important to analyse also which of the parameters can be accepted as common and which differ among different countries, regions or cities.

2 RESEARCH OBJECTIVES

In the study presented preliminary in this paper, the idea was to establish which are the parameters that influence children crossing time defined as time that children spent in area where conflict with vehicles can potentially occure. As observation points, signalized pedestrian crossings near primary schools, were selected. The assumption was that as selected crossings are used by the same children every day, they are relaxed so their instinctive pedestrian behavior can be observed. In order to compare behavior of children of the same age in different traffic environments, the same methodology was applied in Croatia (in the cities of Rijeka and Osijek) and in Italy (in the city of Enna, Sicily). Basic statistical analyses of the collected data were done to establish correlations between crossing time and children caracteristics (gender, age) as well as they behavior (walking in group, running ects.) and traffic condition (crossing length, green pedestrian time) for every analysed case (city) and all analysed parameters.

3 METHODOLOGICAL APPROACH

In this study children-pedestrians were recorded while passing the crossings at signalised intersections. On the basis of recorded passings passing time was measured and defined parameters regarding children notified. Field measurments were made on a total of 18 signalized pedestrian crossings near primary schools, of which 6 in the city of Rijeka, 8 in the city of Osijek and 4 in the city of Enna. Altogether 962 recorded passings were analysed: 331 in Enna, 308 Osijek and 323 in Rijeka. The crossings lengths were from 7 to 16 m, and the width from 2 to 4,5m. At each pedestrian crossing, the duration of total traffic cycle (for vehicles) was measured as well as the duration of pedestrian green light. The passings were recorded in the morning period, between 7.00 and 9.00, when children were expected to arrive to the school. All passings were recorded with camera and later these video clips were analyzed in detail.

From Table 1 it can be seen that in all 3 cities, boys (M) and girls (W) are almost equally represented in the samples. Mostly they cross alone and almost never use mobile phone. More then 50% of children in Enna cross the pedestrian crossing under adult supervision, while those numbers in Croatia are lower. Children are mostly between 8 and 15 years old.

4 RESULTS OF THE RESEACH

Analysis of variance (ANOVA) is one of the most commonly used statistical method to determine statistically significant differences between the three or more data groups (Landau, Everitt, 2004). It enables to determine the impact of individual parameters that are more or less important in the examined phenomenon (Marusteri, Bacarea, 2010). The P value, or calculated probability (Niedoba, Pięta, 2016) is the probability of finding the observed results when the null hypothesis (H0) of a study question is true. The term significance level (α) is used to refer to a pre-chosen probability and the term "P value" is used to indicate a probability. If P value is less than the chosen significance level then null hypothesis (usually "no significant difference between the data groups") is rejected. Table 2 shows P values for different levels of significance in all three observed samples and for all observed input parameters.

The assumption that there is a correlation between input data and dependent variable - time spent in the conflict zone or the time duration of the child movement through the pedestrian crossing [s] was analyzed. The verification the null hypothesis "*All means are equal*" is based on the performing F-test (One-way ANOVA) with significance level $\alpha = 0,05$. The results of the analysis using the ANOVA methodology (Niedoba, Pięta, 2016) are shown in Table 3.

Table 1. Field measured data.

LOCATION	total number of measurments	number of crossings	lenght of crossings [m]	width of crossings [m]	green time for pedestrians [s]	gender	crossing (in a group or alone)	supervized by adults	age of children	talking on mobile phone	using mobile phone (SMS/Internet)
R	323	6	7-16	4	9-13	52% W	64% alone	89% NO	10% <5Y&>15Y	99% NO	97% NO
						48% M	36% group	11% YES	90% 8-15Y	1% YES	3% YES
O	308	8	7-14	2.2-4.5	13-50	47% W	75% alone	94% NO	19% <5Y&>15Y	94% NO	93% NO
						53% M	26% group	6% YES	81% 8-15Y	6% YES	7% YES
E	331	4	9.5-12.5	2-2,5	40	54% W	89% alone	45% NO	35% <5Y&>15Y	94% NO	91% NO
						46% M	11% group	55% YES	65% 8-15Y	6% YES	9% YES

R – Rijeka, O-Osijek, E-Enna

Table 2. Probability of the impact of input parameters.

Input parameters	≤0,001			≤0,01			≤0,05			≤0,1		
	R	O	E	R	O	E	R	O	E	R	O	E
age group			-	-				-				
motoric disabilities		+	+									
movement in a group	+	+	+									
supervision by adults	+		+								+	
mobile – messages								+				
length of pedestrian crossing [m]	+	+	+									
pedestrian green light [s]							+					
run across	-	-										
total number of children on pedestrian crossing		+										
total number of pedestrians		+										

R–Rijeka, O-Osijek, E-Enna; The sign +/- means the positive/negative correlation with the dependent variable

From Tables 2 and 3 it can be seen that the four input parameters on samples of more than 300 measurements of dependent variables (duration of the child movement through the pedestrian crossing) are significant in all three observed environments. Those are "age group", "movement in a group", "supervision by adults" and "length of pedestrian crossing". The input parameters "motoric disabilities" and "run across" have been significant in two of the three environments. Parameters "mobile phone", "duration of green light", "total number of children" and "total number of pedestrians on pedestrian crossing" have been significant only in one of the three environments. Of the 14 input parameters, 4 of them ("gender of a child", "mobile phone – talk", "width of pedestrian crossing", "traffic light cycle duration") according to the P-values, did not meet the lower boundary of significance.

5 CONCLUSIONS

Research that aimed to establish parameters which influence children crossing time at intersections, was simultaneously done in two different countries (Croatia, Italia), and three cities (Enna, Osijek, Rijeka), by observing behavior of children, aged 5-16 years. The results showed that important parameters, when crossing times at signalised intersection is analysed, are: "age group", "movement in a group", "supervision by adults" and "length of pedestrian crossing". Some other parameters like "use of mobile phone", "duration of green light", "total number of children" and "total number of pedestrians on pedestrian crossing" were signicant in just one of the cities but to declare them to be of local significance more detailed analyses in other analysed situations is necessary. In further research the idea is to develop and test unique model for prediction of children-pedetrian time in conflict areas of pedestrian crossings by using neural network approach. Also new locations will be added to the research.

Table 3. The results of ANOVA.

ANOVA	location	Input parameters									
		age group	motoric disabilities	movement in group	supervision by adults	mobile - messages	length of pedes- trian crossing[m]	duration of green light[s]	run across	total number of children	total number of pedestrians
DF	E	6	2	2	1	-	2	-	-	-	-
	R	6	-	4	1	-	3	2	1	-	-
	O	6	1	4	1	1	4	-	1	7	10
SS	E	337,7	218,8	192,8	201,1	-	136,0	-	-	-	-
	R	86,91	-	172,2	95,44	-	895,9	860,7	301,1	-	-
	O	68,61	171,6	278,1	9,47	25,07	382,1	-	137,5	212,4	283,0
MS	E	56,28	109,4	96,4	201,1	-	68,02	-	-	-	-
	R	14,49	-	43,06	95,44	-	298,6	430,4	301,1	-	-
	O	11,43	171,6	69,53	9,47	25,07	95,52	-	137,5	30,34	28,30
F	E	11,00	20,02	17,36	36,52	-	11,82	-	-	-	-
	R	2,86	-	9,04	19,26	-	120,6	166,8	69,77	-	-
	O	2,80	46,71	20,70	2,25	6,04	31,68	-	36,32	8,40	8,30
P	E	0,000	0,000	0,000	0,000	-	0,000	-	-	-	-
	R	0,010	-	0,000	0,000	-	0,000	0,000	0,000	-	-
	O	0,011	0,000	0,000	0,13	0,015	0,000	-	0,000	0,000	0,000

DF – degrees of freedom; SS – Sum of Squares; MS - mean squares; F – F value; P – P value; R–Rijeka, O-Osijek, E-Enna

ACKNOWLEDGMENTS

The research presented in this paper is fully supported by the University of Rijeka under the project "Traffic Infrastructure in the Function of Sustainable Urban Mobility" (uniri-tehnic -18-143).

REFERENCES

Ampofo-Boateng, K.; Thomson, A.J. (1991): Children's perception of safety and danger on the road, British Journal of Psychology, 82, pp 487–505.

Chang, K; Foss, P.; Larrea, Bautista, E.: Student Pedestrian Walking Speeds at Crosswalks Near Schools, (2018), Transportation Research Board, 2018, DOI: 10./1770361198118786814

Gitelman, V.; Levi, S.; Carmel, R.; Korchatov, A.; Hakkert, S.: Exploring patterns of child pedestrain behaviors at urban intersections (2019), Accident Analysis and Prevention, 122, pp36–47 https://doi.org/ 10.1016/j.aap.2018.09.031.

Landau, S., & Everitt, B.:A handbook of statistical analyses using SPSS (Vol. 1). (2004), Boca Raton, FL: Chapman & Hall/CRC.

Marusteri, M., Bacarea, V.: Comparing groups for stistical differences: how to choose the right statistical test?, (2010), Biochemia Medica 2010; 20(1):15–32, https://doi.org/10.11613/BM.2010.004.

Niedoba, T., Pięta, P.: Applications of ANOVA in mineral processing, (2016), Mining Science, vol. 23, 2016, 43–54, doi: 10.5277/msc162304.

Wang, H.; Tan, D.; Schwebel, D.C.; Shi, L.; Miao, L. (2018): Effect of age on children's pedestrian behaviour: Result from an observational study, Transportation Research Part F, 58, 2018 , pp. 556–565, https://doi.org/ 10.1016/j.trf2018.06.039.

Wang, H.; Schwebel, D.C.; Tan, D.; Shi, L.; Miao, L. (2018): Gender differences in children's pedestrian behaviors: Development effects, Journal of Safety Reserach, 2018, 67, pp. 127–133, https://doi.org/ 10.1016/j.jsr.2018.09.003.

World Health Organisation (WHO), 2015. Ten Strategies for Keeping Children Safe on the Road.

A hybrid approach for prioritising road safety interventions in urban areas

S. Vieira Gomes, C. Roque & J.L. Cardoso
Laboratório Nacional de Engenharia Civil, Portugal

ABSTRACT: Spatial organization in cities has often given a special attention to the requirements of motorized vehicles, neglecting pedestrians and cyclists' needs. In order to privilege these road users, there is a need to consider their vulnerability and ensure they are provided with comfortable safety levels.

Pedestrian safety can be improved with the knowledge of the underlying factors involved in crash occurrence and resulting injuries. The consideration of variables related to the built environment in explanation of crashes and injury outcomes is frequent and helpful.

The city of Lisbon was the target for the development of models for the prediction of crash frequencies and their severity, wich could be used to identify high crash risk sites. This procedure contributed to support the intervention in road safety, concerning decisions about the choice of locations for intervention, the characterization of their safety problems, the selection of proper corrective interventions to implement as well as the assessment of the effects obtained with those interventions.

1 INTRODUCTION

Pedestrians are known as the most vulnerable road users, which means their needs and safety require specific attention in strategic plans. In urban areas, where walking is an essential way of travel, this issue may rise to critical levels, since pedestrians are more susceptible to injuries compared to other road users. Portuguese urban areas are no exception: from 2010 to 2016, 97% of the injuries suffered by pedestrians occurred in urban areas.

Walking in a safe environment is indispensable to promote the pedestrian mode. In countries where good results have been achieved with corrective infrastructure intervention, the respective action programs include methods for technically and scientifically rigorous approach to the various stages of intervention. This includes the diagnosis of crashes in the area under consideration, in order to identify the areas where the road environment has a higher influence on crash occurrence.

This was performed for the city of Lisbon, through the development of models for estimating crash frequency, using the Generalized Linear Modelling (GLM) approach with a Poisson-gamma distribution (Vieira Gomes, 2013). The results were improved afterwards through the application of the Empirical Bayes Method (EBM), and used for the detection and hierarquization of high crash risk sites.

More recently, pedestrian severity models were developed also for the city of Lisbon using a Multinomial Logit structure, to identify the main factors involved. Amongst them it was possible to identify key factors related to the urban environment, namely office areas and abrupt driver manoeuvres, which, according to the model, increased the probability of death of pedestrians (Vieira Gomes, et al., 2018).

2 OBJECTIVE OF THE PAPER

This information was combined with the one previously obtained through the crash prediction models in a geocoded database, allowing for a hybrid identification of high crash risk sites.

This tool contributes to support the intervention in road safety in what concerns the identification of promising locations for intervention, the characterization of crashes at each of these locations and their respective safety problems, the selection of the proper corrective interventions to implement, as well as the assessment of the effects obtained with the interventions.

3 METHODOLOGICAL APPROACH

In pursuit for the mitigation of crash occurrence, it is extremely important to integrate the safety aspects in the urban road network planning and management.

Due to the enormous costs of road crashes to society, the knowledge of the factors affecting the likelihood of a crash to occur has been an area of research for many decades. Several researchers have addressed this issue (Lord and Persaud, 2004) reduced it to the understanding of the factors affecting crash occurrence in a given geographic area (usually a segment or intersection) for a certain period of time (week, month, year or group of years). These models consist of mathematical functions that describe the relationship between road safety (expressed in crashes) and explanatory variables, such as traffic, the length of road, the carriageway width, the number of intersections, etc. Such models can be a tool for widespread use, but require, however, an appropriate adaptation to the specific road context.

Several authors have developed crash prediction models adapted to urban areas. In Portugal, Vieira Gomes (2013) developed models to estimate the frequency of crashes for the urban network of Lisbon as a function of diferent road elements, using GLM approaches with the Poisson-gamma distribution. The following disegregations were considered: according to the road element – at intersections (three legs, four legs, and roundabouts) and in segments; and according to the inclusion of explanatory variables related to the road environment – simplified (only with the exposure variables) and global (with all potential explanatory variables).

Poisson and Negative Binomial regression models can be considered the most common statistical methods for crash prediction (Lord *et al.* 2005). However, the factors that influence crash frequency may differ from those that affect crash severity, as stated by (Savolainen *et al.*, 2011).

Having this in mind, another study was developed, specifically aiming to investigate the contributing factors that affect the severity of pedestrian crashes in the city of Lisbon, Portugal. The potential explanatory factors included seasonal attributes, infrastructure characteristics, crash injury outcomes, and information on pedestrian and driver charateristics and manoeuvres. An explanatory modelling approach (based on past observations) was used, with a multinomial logit model (MNL).

4 DESCRIPTION

This chapter presents the two processes adopted in order to identy priority sites for corrective interventions in the city of Lisbon: one through the use of crash prediction models within the Empirical Bayes Method, and the second through the developement of crash severity models.

4.1   *High crash risk sites identification with crash prediction models and the Empirical Bayes Method*

The crash prediction models developed for Lisbon were used to identify high crash risk sites. In practice, these locations can be identified if the expected frequency of crashes is higher than the values calculated by applying the crash prediction models developed for similar entities.

The models for estimating the crash frequency of a type of road element (for e.g. a three-leg roundabout) are developed based on historical data on the observed frequency of crashes in several similar elements, and provide, as such, an expected average value. However, through the application of the Empirical Bayes Method it is possible to improve the estimation of the crash frequency in a specific element (e.g. a roundabout) through a combination of the observed number of crashes with the estimated values (Elvik, 2008).

The most promising high crash risk sites can be identified by calculating the differences between the expected crash frequency of one entity and the expected frequency in similar entities. The cases with larger differences correspond to sites where the crash rate is higher than would be expected, and so they are priority candidates for corrective intervention. Sorting the all similar road elements by descending order of expected crash frequency difference allows to identify priority sites.

4.2   *High risk areas identification with crash severity analysis*

To accomplish this analysis of the factors associated with injury severity levels that pedestrians experienced in the city of Lisbon, a multinomial logit (MNL) model was used to estimate pedestrian and driver injury outcomes, by severity level. Altogether, 19 variables were calibrated and used to identify the potential effects of different factors related to the categories listed above. Significant severity predictors kept in the model included driver injuries, drivers' maneuvers before crash, driver gender, pedestrian age, crosswalk type, land use characteristics, lighting conditions, and time of day.

To further explore the outputs of the modelling process, it is possible to analyse the spatial location of the crashes where they are most likely to do higher damages. Figure 2 presents an example of the location of pedestrian crashes, highlighting the ones that are within the vicinity of businesses areas, which, according to the model output, presents a probability of fatal injury 7.9 times higher than the remaining areas.

Figure 1.   Highest crash risk sites identification (graduation from red - highest, to yellow – lowest).

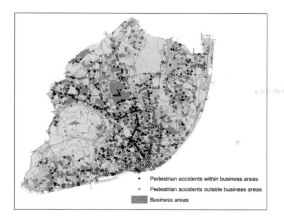

Figure 2.   Pedestrian crashes with higher probability of fatal injury within business areas.

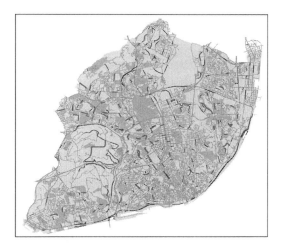

Figure 3.   Hybrid identification of high risk sites for Lisbon.

5   RESULT OF THE RESEARCH

Both models presented priviously have been used to identify high crashes and injuries risk sites. The obtained results were combined, allowing for a hybrid identification and sorting of priority areas for intervention, based not only in crash occurrence, but also in injury reduction.

6   CONCLUSIONS

Trough the combination of crash frequency and severity models developed for the city of Lisbon it was possible to identify high risk sites, using both criteria.

The procedure contributs to support the intervention in road safety, concerning decisions about the choice of places for intervention, the characterization of their safety problems, the selection of the proper corrective interventions to do as well as the assessment of the effects obtained with the interventions.

REFERENCES

Elvik R., (2008). The predictive validity of empirical Bayes estimates of road safety. Accident Analysis and Prevention, 40 (6), pp. 1964–1969.

Lord, D., Persaud, B., (2004). Estimating the safety performance of urban road transportation networks. Em Accident Analysis and Prevention, nº. 36, pág. 609–620.

Lord, D., Washington, S., Ivan, J., (2005). Poisson, Poisson-gamma and zero-inflated regression models of motor vehicle crashes: balancing statistical fit and theory. Accident Analysis and Prevention no. 37, pp. 35–46.

Savolainen, P., Mannering, F., Lord, D., Quddus, M., (2011). The statistical analysis of highway crash-injury severities: a review and assessment of methodological alternatives. Accid. Anal. Prev. 43 (5), 1666–1676.

Vieira Gomes, S., (2013). The influence of the infrastructure characteristics in urban road accidents occurrence. Accident Analysis & Prevention, Volume 60, 289 297.

Vieira Gomes, S.; Roque, C.; Cardoso, J. L., (2018).The severity of pedestrian crashes in Lisbon. 31st ICTCT Conference - On the track of future urban mobility: safety, human factors and technology, Porto.

Pedestrians, Urban Spaces and Health – Tira, Pezzagno & Richiedei (eds)
© 2020 Taylor & Francis Group, London, ISBN 978-0-367-46171-3

Defining the characteristics of walking paths to promote an active ageing

C. Cottrill
Department of Geography and Environment, University of Aberdeen, UK

F. Gaglione, C. Gargiulo & F. Zucaro
Department of Civil, Architectural and Environmental Engineering, University of Naples Federico II, Italy

ABSTRACT: During the next three decades the "grey" segment of population will grow by nearly 80% in many developed countries such as Italy. The structure and the design of outdoor spaces and the related walkable network can improve the quality of life of old population, making it "active" by promoting social engagement and health-based activities.

In this perspective, this paper provides a GIS-based methodology that, starting from the physical characteristics of the pedestrian network and the urban context influencing the choice of a route, is aimed at defining the pedestrian network suitable for the elderly for achieving some of the main activities of their interest.

This goal represents a step of the MOBILAGE project, aiming to define a decision support tool for public administrations to quality of life of elderly.

1   STATE OF THE ART

The trend of worldwide elderly population shows that it is growing at a faster rate than the total population, due to advances in medicine and technology which have led to an increase in life expectancy in the last twenty years (Svensson, 2009). This demographic change poses numerous social, economic, cultural and urban challenges. For this reason, both researchers and local decision-makers are called to work to adapt and reorganize urban spaces to new demand mainly in terms of accessibility and usability (Thakuriah et al., 2011), in accordance with the principles of Universal Design (Hanna, 2005). Infact, the pursuit of Universal Desing is to accommodate the changes that people experience in the course of life, by emphasising on user-centred design (Ginnerup, 2009).

In this perspective, the main research objective of the Mobilage project is to define a decision support tool for public administrations, in order to increase pedestrian accessibility of urban services for the population over 65. Within the Mobilage project, the Federico II research unit has been working to innovate, also from the methodological point of view, the traditional urban planning "trade tools" to adapt them to the new socio-cultural needs. More in detail, the research group worked at the definition of Functional Accessibility Soft zones which want to represent an innovation of the by now obsolete concept of the traditional catchment area and can constitute a starting element to reorganize the urban services to better satisfy the renewed needs of population. The Mobilage project develops two main integrated research steps that have the common goal of identifying the Functional Accessibility Soft zones for services of interest for the over 65 population: (i) the definition of urban portions where services of interest are located, considering their distribution and according to the behavior of the over 65s; (ii) a GIS-based methodology that, starting from the physical characteristics of the pedestrian network (eg the presence of sidewalks) and the urban context's (eg proximity of green areas) influencing the choice of a route, defines what makes a pedestrian network more suitable for the elderly to achieve their activities of interest. This last research segment is the main topic of the paper.

The improvement of pedestrian accessibility (walkability) for the elderly to urban services is one of the most important elements to improve the overall organization of the urban system, as well as to reduce the social exclusion of this specific segment of population (Tiboni and Rossetti, 2014; Khosravi et al, 2015; Gargiulo et al., 2018). In fact, one of the main themes of interest for the scientific community refers to the improvement; of the pedestrian network for the over 65s, considering that soft mobility is the most widely used mode of travel by the elderly (Banister and Bowling, 2004; Eurostat, 2015).

In summary, there are two main lines of research on this issue: the first aims to define the physical and environmental characteristics of a pedestrian path in order to improve soft mobility (Negron-Poblete et al., 2016; Wang et al., 2019); the second is mainly oriented to the development of a synthetic qualitative or quantitative indicator of "walkability" (Svensson, 2009; Loh et al., 2019).

Therefore, research and studies have shown that urban spaces and pedestrian paths are mainly studied either in terms of organizing the soft mobility networks or in relation to the habits of the elderly. There are few studies that, considering the identification of pedestrian paths suitable for the elderly, also take into consideration the location of the activities and services of interest for them.

2 OBJECTIVE OF THE PAPER

The work described in these pages pursues a twofold objective: (i) identifying the pedestrian paths suitable for the elderly based on their physical urban context and the safety characteristics, and (ii) defining the Functional Accessibility Soft zones (FAS zones) on the basis of the "optimal" pedestrian routes to reach urban services of interest for elderly.

In this perspective, this study aims at developing a methodology for the definition of Functional Accessibility Soft zones (FAS zones) that identify the portions of the municipal territory characterized by the presence of pedestrian paths suitable for the elderly that facilitate the achievement of services of interest for this segment of population. To identify the FAS zones the behaviors of the over 65s have been related to the supply of main services through the network of pedestrian routes, in order to better respond to their demand for accessibility taking into account the principles of Universal Design and the most recent theories on accessibility (Levison and Lahoorpoor, 2012).

3 METHODOLOGY

The proposed methodology is divided into four phases. The first phase concerned the classification of the set of variables useful for defining the network of pedestrian routes, according to the following three categories:

1. physical characteristics that refer to the geometry and quality of pedestrian routes;
2. characteristics related to the sense of security that refer to the perceived protection in walking a pedestrian path;
3. characteristics of the urban context that refer to the attractiveness and usability of a pedestrian path.

Starting from scientific literature, the most significant variables were assigned to each group, following multivariate statistical analyzes, and other variables deemed relevant to the research objectives have also been introduced. The width of the pavement was also considered among the physical characteristics; among the characteristics linked to the sense of security, the density of public lighting was added; for what concerns the urban context, the presence of panoramic points and non-main roads were considered. In particular, the exclusion of the main road network is linked to the fact that air and noise pollution, traffic volumes and

Table 1. Characteristics of the pedestrian network.

ID	Variable	Measure	Source
Physical characteristics			
1	Slope of the links of the road network	>5%=0 o <5%=1	GIS
2	Sidewalk width	< 1,5 m=0 >1,5m=1	Google Maps
3	State of pavement of the sidewalk	0=poor good=1	Google Maps
Characteristics releted to the sense of security			
4	Lighting density	<0,056=0 >0,056=1	GIS
5	Presence of escalators and elevators	No=0 Yes=1	Google Maps
6	Presence of parking areas	No=0 Yes =1	Google Maps
Urban context characteristics			
7	Presence of green areas	No=0 Yes =1	Google Maps
8	Presence of panoramic points	No=0 Yes =1	Google Maps
9	No-main roads	No=0 Yes =1	Google Maps

vehicle travel speeds are generally so high that they are usually perceived as unsafe and unpleasant for pedestrian movements, especially for elderly.

Among all the variables, the only one linked to the context is the density of public lighting, defined as the ratio between number of poles and the length of the arches of the pedestrian network.

In phase two, all the characteristics were associated in the GIS environment to each arc of the pedestrian graph in order to determine the overall qualitative "weight" and, through the natural breaks method, the links of the pedestrian network were classified according to the score obtained for each of the three categories identified (physical characteristics, characteristics linked to the sense of security and urban context), and overall.

In phase three the compatible paths to elderly pedestrian mobility attitude have been identified, starting from the presence of all the features listed above. In particular, this last compatibility was identified on the basis of the following three elements: length of the route based on the elapsed walking time; weighted average of the characteristics associated with each arc; identification of "optimal" paths for the elderly to achieve urban services.

In phase four, referring to the part of research developed before this work and relating to the types of services of interest (see section 1), the FASZones have been identified, which are the areas characterized by the presence of channels suitable to be traveled by the elderly, due to their behavior (speed of movement on foot) and the characteristics of usability and attractiveness that characterize these links of the pedestrian network. To this end, the Network Analysis was used by assigning to each link of the pedestrian graph the value of the weighted average of their characteristics.

4 RESULTS OF THE RESEACH

The methodology described was applied to the V Municipality of Vomero and Arenella of the city of Naples. The choice of this area of study is linked to the demographic structure, characterized by the presence of a population over 75 above the urban average (6000 over 75 compared to an average value of 3400) (Città di Napoli, 2016). By way of example, the results obtained from the application of the methodology refer to the over 75 and to a single service of interest for the elderly, which are pharmacies. The results obtained allow for:

- identify the portions of the pedestrian network in which to intervene primarily to improve both the individual characteristics considered and the usability and overall attractiveness (Figure 1). In particular, the area of Piazza Vanvitelli and Piazza Medaglie d'Oro are characterized by the presence of pedestrian paths suitable for the elderly, unlike the Rione Alto area and the Camaldoli area;

- identify the areas where the users of a service reside, considering the presence of the paths which are perceived as "optimal", possessing all the qualities (characteristics) considered. From Figure 2 it is possible to notice the potential size of FAS zones when all the pedestrian paths are equipped (green-coulored areas), the result of the first Mobilage work step (Gargiulo et al., 2019), and the size of the FAS zones built on the basis of the existing "optimal" routes (yellow areas). The comparison between these FAS zones, therefore, allows to identify the gap, in terms of areas, on which the public decision-maker must intervene to improve accessibility to urban services for the elderly.

Hence, these results provide to local decision-makers a support tool for deciding where and how to intervene both on the pedestrian network and on the distribution and location of services, in order to increase accessibility for the over 75 by improving their quality of life.

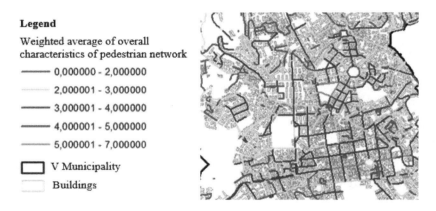

Figure 1. Extracts of the pedestrian paths classified according to the characteristics considered.

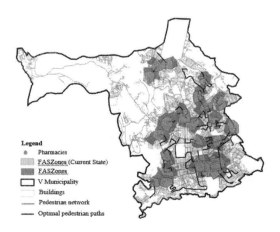

Figure 2. Pedestrian accessibility to pharmacies for the elderly.

AUTHORS ATTRIBUTION

sections 1 and 2 were equally written by C.C. and C.G.; section 3 F.G.; section 4 F.Z.

REFERENCES

Banister, D., and Bowling, A. (2004). Quality of life for the elderly: the transport dimension. Transport policy, 11(2),105–115.

Comune di Napoli (2016). Città al censimento 2011. Accessed 28 October 2018. Available at: http://www.comune.napoli.it.

EUROSTAT 2017. People in the EU - statistics on an ageing society. Accessed 18 December 2018. Available at: https://ec.europa.eu/eurostat/.

Gargiulo C., Zucaro F., and Gaglione F. (2018). A Set of Variables for the Elderly Accessibility in Urban Areas. Tema. Journal of Land Use, Mobility and Environment, 0, 53–66.

Gargiulo C., Zucaro F., Gaglione F. and Faga L. (2019, in press). Improving accessibility to urban services for over 65: a GIS-supported method. In C. Gargiulo & C. Zoppi (Eds.), Planning, nature and ecosystem services. Naples: FedOAPress. ISBN:978–88–6887–054–6.

Ginnerup, S. (2009). Achieving full participation through Universal Design. Strasbourg: Council of Europe Publications.

Hanna, E. I. (2005). Inclusive design for maximum accessibility: A practical approach to universal design. PEM Research Report.

Khosravi, H., Gharai, F., and Taghavi, S. 2015. The impact of local built environment attributes on the elderly sociability, 25(1),21–30.

Levinson, D. M., & Lahoorpoor, B. *(2019). Catchment if you can: The effect of station entrance and exit locations on accessibility.*

Loh, V. H., Rachele, J. N., Brown, W. J., Ghani, F., Washington, S., & Turrell, G. (2019). The potential for walkability to narrow neighbourhood socioeconomic inequalities in physical function: A case study of middle-aged to older adults in Brisbane, Australia. Health & place, 56, 99–105.

Negron-Poblete P., Séguin A. M. and Apparicio P. (2016). Improving walkability for seniors through accessibility to food stores: a study of three areas of Greater Montreal, Journal of Urbanism: International Research on Placemaking and Urban Sustainability, 9:1, 51–72.

Svensson, J. (2009, November). Accessibility in the urban environment for citizens with impairments: using gis to map and measure accessibility in Swedish cities. In Proceedings of the 24th International Cartographic Conference, Santiago, Chile (pp. 15–21).

Thakuriah, P., Sööt, S., Cottrill, C., Tilahun, N., Blaise, T., & Vassilakis, W. (2011). Integrated and continuing transportation services for seniors: case studies of new freedom program. Transportation research record, 2265(1),161–169.

Tiboni, M., and Rossetti, S. (2014). Achieving People Friendly Accessibility. Key Concepts and a Case Study Overview. TeMA Journal of Land Use, Mobility and Environment,.

Wang, R., Lu, Y., Zhang, J., Liu, P., Yao, Y., & Liu, Y. (2019). The relationship between visual enclosure for neighbourhood street walkability and elders' mental health in China: Using street view images. Journal of Transport & Health, 13, 90–102.

Effects of elderly people's walking difficulty on concerns and anxiety while walking on roads

T. Matsuura

Jissen Women's University, Japan

ABSTRACT: The purpose of the study is to examine concerns and anxieties of elderly people while walking on roads. Whether these anxieties and concerns differ according to walking difficulty is also investigated. A total of 342 elderly people from 60 to 87 years-old who are registered in two human resource dispatch agencies participated in the questionnaire. The questionnaire included scales of concerns and anxiety while walking on roads, visual and physical walking difficulty, and daily walking behavior. Both visual and physical walking difficulties increased anxiety about walking conditions. They also increased concerns about path conditions and convenient route. Older people with difficulty seemed to need better conditions and felt anxiety about bad conditions because of their smaller attentional resource to the conditions. In spite of this, people with physical difficulty did not assess safety as more important than those without physical difficulty. This may increase their involvement in pedestrian accidents.

1 INTRODUCTION

Social trends like the growth of car use, aging of the population, and increase in health and environmental problems have caused some studies on walkability and traffic safety in the area. Walking needs or conditions of importance for pedestrians to walk are one topic of such studies. The common conditions included feasibility, accessibility or convenience, safety and security, comfort, and attractiveness or pleasurability (Alfonzo, 2005; Fortyth, 2015; Methorst, 2007). Feasibility seems to be most fundamental of these conditions. However health problems make walking less feasible for older persons. A study showed that older people wanted safety (i.e., a smooth surface on the sidewalk, pedestrian crossings, and signalized crossings) as a fundamental condition more than younger people (Bernhoft & Carstensen, 2008).

Environmental barriers or perceived problems when walking on roads have been also studied. The topic refers to how people feel anxious about various walking conditions. These study were conducted because people were not necessarily satisfied with the local conditions. In other words, environmental barriers make walking activity difficult for people, especially for older people (Hovbrandt et al, 2007). Past studies identified perceived barriers such as bicycles ridden on paths and on sidewalks, heavy and fast traffic, problems with crossing streets, such path conditions as uneven or slippery surfaces, bad lighting, and fear of crime. The barriers reported most often were traffic-related ones like bicycles on sidewalks and heavy traffic (Hovbrandt et al., 2007; Takeshima, 2007).

Older people who find walking somewhat difficult suffer from the barriers more than those who do not. They are concerned about demanding environmental conditions because they must allocate more attention to watching their steps and negotiating obstacles on roads. If the environmental load is high, walking becomes a dual task for them and they may not cope well with the situation (Neider et al., 2011). Thus, they will not only feel anxious to demanding environments but also evaluate walking needs higher than other older people without walking difficulty. Research on walking difficulty focussed mainly on gait disorders, which may result in falls. In older age, gait disorders typically have several causes, which include physical and visual impairment (Pirker & Katzenschlager, 2017).

The aims of this study are as follows: 1) Examine elderly people's concerns or conditions of importance while walking on roads. 2) Examine elderly people's anxiety or barriers while walking on roads. 3) Reveal to what extent elderly people experience visual and physical difficulties. 4) Investigate whether the visual and physical walking difficulties have effects on the walking concerns and anxiety as well as walking behaviors.

2 METHOD

2.1 *Participants and procedures*

Participants were 342 community-dwelling elderly people who were registered in two human resource dispatch agencies. Their ages ranged from 60 to 87 years (mean 71, standard deviation 5.0, female 37%). To recruit participants, the two local human resource dispatch agencies called 'Silver Human Resource Centers' were selected through the central office in Tokyo. Each local office then recruited participants registered there by distributing a questionnaire at some meetings and when members came to the office.

2.2 *Questionnaire*

Concerns while walking were assessed by four scales, each with 3 items; safety (do not be hit by a car, cross the road safely, do not be collided with a bicycle), path conditions (sidewalks are even and no step, sidewalks are wide, free from obstacle such as signboards on sidewalks), pleasant environment (can enjoy scenery of nature, street is lively and attractive, way is quiet), and convenient routes (destinations are close enough to get to, get to a destination in a reasonable time, do not make a detour). The scales were selected in reference to previous studies that indicated main pedestrian needs (Alfonzo, 2005; Methorst, 2007) and our preliminary surveys. Participants answered to the question of 'To what extent are you concerned about the next environmental conditions or situations?' They indicated the degree of concern on a four-point scale (1 = 'not at all' to 4 = 'very') for each of the 12 items.

Anxiety about walking in environmental conditions or situations was also measured using four scales, each with 3 items; crossing streets (crossing a signalized intersection, crossing a street with a push-button signal for walkers, and crossing a street without a pedestrian crossing), path conditions (uneven surface and steps on the sidewalk, narrow sidewalks, obstacles such as signboards on the sidewalk), vehicular traffic (walking on the curb in heavy traffic, walking on the curb in fast traffic, and bicycles on the sidewalk), and bad weather (walking in the evening and at night, walking in bad weather such as rain, walking in hot or cold weather). These scales were selected in reference to our preliminary surveys. Respondents indicated the degree of anxiety on a four-point scale, 1 = 'not at all' to 4 = 'very'.

Visual walking difficulty was measured with three items by asking: 'To what extent do you have difficulty in reading road signs or name of stores, going down steps or curbs in dim light or at night, and noticing objects off to the side while you are walking along?' The items were selected with reference to the Japanese version of NEI-VFQ 25 (Suzukamo et al., 2005). The answer to each item was given on a four-point scale (1 = 'not at all' to 4 = 'always'). A score of visual walking difficulty (3 to 12) was calculated by adding the scores of the three items. Similarly, physical walking difficulty was also measured using three items by asking: 'How often do you experience the following three symptoms: a pain and numbness in feet, legs and a waist, shortness of breath, a heartbeat and chest pain, and leg fatigue due to a drop of muscle strength ?

Lastly, daily walking duration on roads (1 = 'within 15 minutes to 4 = 'more than one hour') and possible walking duration without rest (1 = 'within 5 minutes' to 6 = 'more than one hour') were assessed as walking behaviors of elderly persons.

3 RESULTS

3.1 *Concerns and anxiety while walking on roads*

The concerns differed based on walking conditions. While most pedestrians rated 1 (= not concerned at all), they rated 4 (= very concerned) on three safety items. Older pedestrians found safety most important for walking among four scales of concerns. ANOVA indicated that environmental conditions significantly predicted pedestrian's concerns (F (3,957) = 129.53, p <.001, η^2 = .015). Scheffe's multiple comparison test showed the significant differences between safety and the other three concerns. Internal consistencies measured by Cronbach's alpha for each scale were as high (.78 to .92) as other scales used in the study. Correlations between four concerns showed high correlation coefficients between r = .43 and .68. Elderly pedestrians who regard a condition as important were more likely to regard other conditions as important.

Participants rated 2 (= slightly anxious) on most items. Means of anxiety raged from 1.60 to 2.87. Thus they felt slightly or somewhat anxious about walking conditions. The means of the scales significantly differed (F (3,964) = 185.47, p <.001, η^2 = .017). Scheffe's multiple comparison test showed high anxiety in the following orders; vehicular traffic (M = 2.7), bad weather (M = 2.2), path conditions (M = 2.1), and crossing streets (M = 1.8). The correlations between four scales ranged from .61 to .74 .

3.2 *Walking difficulty relating to walking concerns, anxiety and behaviors*

About half of elderly pedestrians did not report walking difficulty at all in many situations. However, a higher percentage (about 75%) of them reported walking difficulty in situations of 'going down steps or curves (= 2.0)' and 'shortness of breath (=2.0)'. Both scales of waking difficulty correlated significantly with each other (r = .51).

Small to moderate correlations were found between visual walking difficulty and concerns while walking (Table I). However, physical walking difficulty did not correlate with concerns of safety or pleasant environments. Moderate to high correlations between both walking difficulties and anxiety while walking were found (right of Table 1). The correlations with anxiety were also higher than those with concerns. Lastly, visual difficulty had higher relationships with concerns and anxiety than physical difficulty.

Both visual and physical walking difficulty had slightly predicted daily walking duration (r = -.15 and -.19) and moderately predicted possible walking duration without rest (r = -.27 and -.37). Contrary to the correlational results for walking concerns and for anxiety, the correlations were higher for physical difficulty than for visual difficulty.

Table 1. Correlations between walking difficulty and concerns and anxiety while walking.

Walking difficulty	Concerns				Anxiety			
	Safety	Path conditions	Pleasant environment	Convenient routes	Crossing streets	Conditions of path	Vehicular traffic	Bad weather
Visual difficulty	.22**	.38**	.25**	.30**	.54**	.61**	.51**	.57**
Physical difficulty	.09	.23**	.05	.28**	.40**	.44**	.34**	.37**

*p < .05, **p<.01

216

4 DISCUSSION

The study demonstrated that elderly persons generally consider environmental conditions slightly or moderately important when they walk. They found safety most important among four concerns. This result agree with previous research that older pedestrians found safety more important than younger pedestrians did (Bernhoht & Carstensen, 2008). Another study that interviewed older people during their walk also indicated that traffic safety (e.g. busy traffic) and walking facility (e.g. sidewalk quality) were major issues (Cauwenberg et al., 2012).

Participants also felt slightly or somewhat anxious about walking conditions. Most anxious-producing conditions were vehicular traffic including heavy and fast traffic and bicycles on the sidewalk. This result was consistent with previous researches (Hovbrandt et al, 2007; Takeshima, 2007). The study participants did not feel anxiety as much when crossing streets as other environmental conditions. This result contradicts the national pedestrian accident statistics that 56% of them occurs when crossing roads (ITARDA, 2019a).

The results indicated that walking difficulty had great effects on anxiety, and somewhat on concerns and behaviors. A new finding is that walking difficulty affected walking concerns, especially concerns about path conditions and convenient routes. This can be explained by understanding that elderly people with walking difficulty are sensitive to the conditions and need good conditions to compensate for their decreased attentional resources which instead are used mainly to maintain stable mobility on foot (Aviner, Shinar, & Susilo, 2012).

Other findings include the difference between visual and physical walking difficulties. First, walking difficulty due to physical deficits had more effect on daily walking duration and possible walking duration without rest than visual difficulty. While older people with some physical deficits like a foot pain may not continue to walk physically, those with some visual deficits may continue to walk if they attempt it. Second, walking difficulty due to visual deficits had a greater effect on walking concerns and anxiety. A possible reason is that the tasks involved in walking are highly visible; the ability to walk becomes more compromised with impaired vision than with physical deficits. Another reason is that although older people with impaired vision cannot adjust their walking behavior if environmental conditions is very high (e.g. to cross a street unsafely under dark street lighting), those with physical impairment can cope with the bad environmental conditions (e.g. to cross the street after waiting for a while and walk slowly following their pace).

According to road accident statistics in Japan, the percentage of pedestrian accidents while crossing roads is higher for older people (59% for 65 years and over) than for younger one (51% for 20 to 64 years) (IRTADb, 2019). One reason is older people's decreased motor abilities, such as walking speed when crossing the street (Aviner, Shinar, & Susilo, 2012; Dommes, Cavallo, & Oxley, 2013). Nevertheless, elderly people with physical walking difficulty did not assess safety as more important than those without physical difficulty. This may increase their involvement in pedestrian accidents.

5 CONCLUSION

The study has confirmed that a safe environment is most important for elderly pedestrians. It has also confirmed that elderly people with some walking difficulty require better environmental conditions than those without walking difficulty. These findings are important for the development of research and for making paths and sidewalks more walkable. Conditions of importance while walking on roads change according to trip purpose and the routes on which pedestrians walk. How these specific concerns differ from general concerns is a topic to be researched in the near future.

REFERENCES

Alfonzo, M.A. (2005) To walk or not to walk? The hierarchy of walking needs. Environment and Behavior. 37 (6), 808–836.

Avineri, E., Shinar, D., & Susilo, Y.O. (2012). Pedestrians' behaviour in cross walks: The effects of fear of falling and age. Accident Analysis and Prevention, 44, 30–34.

Bernhoft, I.M., & Carstensen, G. (2008). Preferences and behavior of pedestrians and cyclists by age and gender. Transportation Research F, 11, 83–95.

Cauwenberg, J.V., Holle, V.V., Simons, D., et al. (2012). Environmental factors influencing older adults' walking for transportation: a study using walk-along interviews. International Journal of Behavioral Nutrition and Physical Activity 9, Article number: 85.

Dommes, A., Cavallo, A., & Oxley, J.A. (2013). Functional declines as predictors of risky street-crossing decisions in older pedestrians. Accident Analysis and Prevention, 59, 135–43.

Forsyth, A. (2015). What is a walkable place? The walkability debate in urban design. Urban Design International, 20, 274–292.

Hovbrandt, P., Ståhl, A., Iwarsson, S., Horstmann, V., & Carlsson, G. (2007). Very old people's use of the pedestrian environment: functional limitations, frequency of activity and environmental demands. European Journal of Ageing, 4(4),201–211.

ITARDA (2019a). Traffic Statistics 2018 Edition (In Japanese). Institute for Traffic Accident Research and Data Analysis.

ITARDA (2019b). Cross-tables of road accident, 30-13NM202 (In Japanese). http://www.itarda.or.jp/materials/statistical.php

Methorst, R. (2007). *Assessing pedestrians' needs. The European COST 358 PQN Project*. https://www.researchgate.net/publication/240621797_Assessing_Pedestrians'_Needs_The_European_COST_358_PQN_project.

Neider, M.B., Gaspar, J.G., McCarley, J.S., Crowell, J.A., Kaczmarski, H., Kramer, A.F. (2011). Walking and talking: dual-task effects on street crossing behavior in older adults. Psychology and Aging, 26(2), 260–268.

Pirker, W., & Katzenschlager, R. (2017). Gait disorders in adults and the elderly. Wien Klin Wochenschr, 129, 81–95.

Suzukamo, Y., et al. (2005). Psychometric properties of the 25-item National Eye Institute Visual Function Questionnaire (NEI VFQ-25), Japanese version. Health and Quality of Life Outcomes, 3, Article number: 65.

Takeshima, Y. (2007). Research on the walk environment maintenance for the aged: The order of the barrier and how to cope those barriers in the daily going out. Journal of architecture, planning and environmental engineering, 611, 1–6 (In Japanese).

Healthy cities for all

Pedestrians, Urban Spaces and Health – Tira, Pezzagno & Richiedei (eds)
© 2020 Taylor & Francis Group, London, ISBN 978-0-367-46171-3

Shared space and visually impaired persons

F. Anwar
Università degli Studi di Brescia, Italy

ABSTRACT: Visually impaired populations have been widely ignored in the past. This paper focuses on the problems that visibly impaired have while being in a shared space. The movement of persons in a shared space depends a lot on the sense of sight, as it's a single space used by different users with no barriers.The issue has attracted much debate and discussion. This work shows how persons with visual impairment can use the shared space in a better and safer way. I choose Via Garibaldi in Turin, Italy for my fieldwork using different methods through the site visits individually and with the user/experts which helped me in understanding that how they co- exist with the other users in a shared space and what problems they had while walking. The results attained helped me to formulate the suggestions to improve the shared space of Via Garibaldi for visually impaired persons.

1 INTRODUCTION

In the past the movement of visually impaired persons in urban space has been widely ignored and they have been facing great difficulties while moving alone in an urban environment. Shared space is a new design concept for town centre and high street developments. It mostly involves removing the kerb that has traditionally separated areas for vehicles and pedestrians creating a common area for all the users and mixing them together. The concept is to make the street more social and people friendly .The idea was to become acquainted with appropriate terminology and models to understand the experiences of persons with impairment and come up with some suggestions that can help to improve the shared space for visually impaired persons.

2 OBJECTIVE OF THE PAPER

In this project, I discussed with examples how persons with a visual impairment experience the built environment by their senses. After going through the literature I came up with two main questions for our research

How does shared space works for visually impaired persons?

How to use other senses apart from the sense of sight in design?

3 METHODOLOGICAL APPROACH

I used different methods through the site visits individually and with the user/expert. I visited the site with three user/experts at different times of the day. Two of them were completely blind while one of them could see from one eye a bit. The visit of Via Garibaldi with the user/

experts helped me in understanding that how they co-exist with the other users in a shared space and what problems they had while walking. Understanding the problems that they face while being in a shared space and urban context. Visiting the place at different times and at different rush hours was very helpful to understand how they manage to move. This user/ expert analysis was very useful to figure out the different hindrances that visually impaired persons faced in Via Garibaldi. I was able to understand different obstructions and how to remove them and also to analyze that the problems of completely blind persons are very different from partially impaired persons. The individual visits helped me a lot . It gave me another perspective of the street as I imagined it walking the street as a visually impaired person and tried to figure out the problems.

There were certain points that I focused on:

3.1 *People selling goods, the cars, road crossings, the traffic signals, the pavements*

In the user/expert analysis the expert users are involved who can give you the best feed back as they are the experts and can point out the things that other normal users can not. The pace is understood and conceived in a better way. The persons who are blind are able to appreciate sounds, smells or haptic qualities that designers may not be attuned to. Everyone experiences space in a multisensory way, yet for most people sight tends to receive more attention than the other senses. Because blind persons and especially persons without residual vision are as it were experts in non-visual sensory perceptions, they seem ideally placed to uncover the multisensory qualities of a building or space.

I managed to gather three volunteers and observed the movement of the expert/user through, Interviews, Making notes, Videos, Pictures. I visited the street with each of them at two different times of the day as at different times there are different users and difference in traffic. With the first person I started the first visit at 1pm he mentioned that there are more people in the street because of lunch break. The second time at 3pm he mentioned that there are less people in the street because the people are working in offices and shops.

4 RESULT OF THE RESEACH

After collecting all the data through the user/expert analysis the next step was to study the data and bring above the points which were pointed by the expert/users . Mainly the defects and then give suggestions to tackle these defects. The focus was on: Sign boards, Obstacles, Reference sounds, Traffic signals, Walking strategy, Pedestrians Cars, Bicycles and Street vendors.

There were some commonalities and differences between completely blind and partially blind persons:

Table 1. Some commonalities and differences noticed between the partially and completely blind persons user/expert analysis.

COMMONALITIES	DIFFERENCES
Walking strategy	Color contrast
use of sounds	Use of stick
	Knowledge of the street

Following are some of the solutions that I thought are necessary keeping in mind the points coming out from the user/expert analysis in Via Garibaldi.

Table 2. The solution and description for the shared space of Via Garibaldi.

SOLUTIONS	DESCRIPTION
1-Tactile pavements	The visually impaired persons suggested that the street lacked a tactile pavement which was very necessary. My idea is to use a small strip which would give the concept of safe space. This strip will be very small just for the visually impaired persons as a warning sign .The Safe Space would be seen as the equivalent of the footway in a traditional street but it would not prevent motorists, cyclists and pedestrians from sharing the larger part of the street area - the shared space – where they are confident about doing so. In this approach the shared space area would occupy only a part of the street - usually the center part. Also on the road crossings a warning strip is necessary for the visually impaired persons.
2-Bicycles	Cyclists will want to park as near as possible to their destination, typically less than 50 m. As there are no bike stands so I suggest the induction of stands next to the bikes lane along with the benches. This demand needs to be met with a dispersed offer of small parking facilities, at short distance intervals In this way the bikes stay in the middle of the street on each side .
3-Traffic signals	The Audio Tactile Pedestrian Detector has a vibrating button which emits both audio and tactile signals indicating that the green man signal is lit and that it is therefore safe to cross the road. The device is built into the pole, and is used throughout the world at road crossings that blind and deaf people use.
4-Soundscape	Information about the environment, such as the height and width ratio of the street, can be derived from the reflection of the sound of a long cane or the contact of one's foot with the ground. In this way one notices that a building is close by, or that a building line ends. The introduction of small water bodies and greenery can be used as non-visual landmarks by the visually impaired persons .The smell of water and plants can be used as an indication of a safety zone
5-Comfort zone	It may be necessary to preserve this natural division explicitly in the design by means of a comfort zone where pedestrians and other vulnerable road users are given space and where no other traffic will be expected. a comfort zone can be achieved by using different colors and materials, and by the installation of street furniture .
6- Brightness and color contrast	Clear color contrasts can be used to indicate a difference in function. When marking a walking route or pedestrian crossing it is particularly important that the difference in brightness between materials and colors is sufficient. Street lighting must be sufficiently clear and even and should be activated at appropriate times.
7-Displays	A walking route along a façade line requires that displays and terraces are arranged in such a way that sufficient walking space is left. The façade is free of obstacles and forms a continuous line where possible.

Figure 1.   Shared space and safe space.

5   CONCLUSIONS

A final aspect, on which I want to shed light, is the possible further research. Through this research we have discovered there is a lot of literature about the experiences of visually impaired persons, the accessibility regulation especially focuses on the needs for physically impaired persons. However, the user/experts noticed a lot of obstacles in the shared space.

In future research more stake holders can become part of the user/expert analysis and more better solutions can be found to make shared space more safer. There is a lot of room to find ways to educate people about the shared space. Hopefully in near future more studies will be done on this. The proposed changes can have a positive effect on the overall look of the Via Garibaldi and not only the visually impaired persons will admire the changes but also the other users of the street will be able to move in a better and safer way.

REFERENCES

Hamilton-Baillie, B. (2008a). 'Shared Space: Reconciling People, Places and Traffic' .Built Environment, 34(2), p161–181.
Hamilton-Baillie, B. (2008b). 'Towards shared space'. Urban Design International, 13,p130–138.
Heylighen, Ann, 2010 et al. "Challenging prevailing boundaries in design", in: Boederline, Pushing design over the limit, Cumulus.
Imrie, Rob, 2003 (21) "Architects' conceptions of the human body", in: Environment and Planning D: Society and space,, pp. 47–65.
Matteo Schianchi,, 2009 "La terza nazione del mondo. I disabili tra pregiudizio e realtà", Feltrinelli, 11.
Rob Methorst (2003): Vulnerable Road Users, Adviesdienst Verkeer en Vervoer, Rotterdam.
Shared Space. (2005). Room for Everyone: A new vision for public spaces. Leeuwarden:Shared Space.
Thomas, C. (2007, 14 Sept). 'Carol's Blog: Blinding rules of the road'. NCE. Retrieved from http://www. nce.co.uk/carols-blog-blinding-rules-of-the-road/85886.

Pedestrians, Urban Spaces and Health – Tira, Pezzagno & Richiedei (eds)
© *2020 Taylor & Francis Group, London, ISBN 978-0-367-46171-3*

Topographical and physiological data collection for urban handbike tracks design

A. Cudicio, A. Girardello, F. Negro & C. Orizio
Department of Clinical and Experimental Science, University of Brescia, Italy

A. Arenghi
Department of Civil, Enviromental, Architectural Engineering and Mathematics, University of Brescia, Italy

G. Legnani
Department of Mechanical and Industrial Engineering, University of Brescia, Italy

M. Serpelloni
Department of Information Engineering, University of Brescia, Italy

ABSTRACT: Cities should guarantee, to all the citizens, spaces for physical activity to allow people to reach the correct amount of activity during the week. Physical activity areas for people with disability need specific safety and accessibility precautions. The aim of this study is to give suggestion to modify tracks suitable for handbike practice and to classify and describe them in order to give useful information to the users. Ten healthy subjects were tested on a dedicated handbike equipped with a powermeter. The oxygen consumption, heart rate, speed, track elevation profile and distance during the execution of one selected track were simultaneously detected. This experimental set up resulted a reliable tool to describe the relation between the track topographical features and the individual metabolic engagement as a function of the speed. Further measures, on subjects with motor disability, will be the crucial translational phase of the whole project.

1 INTRODUCTION/STATE OF THE ART

Physical activity improves physical fitness in all its components, mood and socialization (Barros et al. 2017; Donnelly et al. 2017; Ekelund et al. 2018). In particular, in a population with reduced mobility the execution of regular adapted physical activity (APA) can improve lifestyle quality, prevent the rising of non-communicable diseases and contribute to increase the expectancy of active life. Inaccessibility of structures or places for adapted physical activity practice must be overcome by institutional urban policy (Rimmer et al. 2004). Indeed, even the availability of tracks for hand-biking practice within urban areas is scarce. In particular, the identification of enjoyable tracks in green areas is lacking in most of the western cities.

According to the literature (CDC.GOV/disabilities/PA) the practice of APA needs both accessibility of the exercise/sport facilities and a tailored exercise prescription based on individual physical fitness in relation to the track characteristics. Currently, 57% of the adults with motor disability does not reach the recommended amount of aerobic physical activity.

Our group of research would contribute to face the above reported issues supporting the design of useful tracks in town for handbikers, mostly represented by persons with spinal cord injury.

2 OBJECTIVE OF THE PAPER

In this view this project is aimed to identify a method based on the simultaneous recording of ergometabolic, kinematic and topographical data in order to: 1. provide the Brescia municipality with useful and applicable suggestions to modify or adapt the already existing paths and make them suitable for handbike use; 2. classify the different tracks on the basis of the required external work and let the users choose the best pattern according to their fitness.

3 METHODOLOGICAL APPROACH AND/OR RESEARCH APPROACH

The first step of this project was the identification of the tracks suitable for handbike use; there are several places in Brescia where it is possible to cycle, but handbike needs tracks with peculiar features. First of all, the route should be easy to be reached by public transportation or by car, with dedicate parking lot. Then, the surface has to be regular, without big potholes or slippery surfaces ensuring a safe and efficient cycling. Handbikers need a lay-by zone where to rest without occluding the street and large enough to allow the change of direction. Finally, the tracks must not include dangerous parts both for the bikers and for the other people (for example pedestrian crossings or busy intersections).

Once identified the tracks (an example is reported in (Figure 1)) they will be studied according to their features such as the elevation profile, the surface and the requested ergometabolic engagement of the subject. Eventually, the following factors will be measured and reported:

Figure 1. Example of track (red line) in a public and car free area.

altitude variation along the path with respect to the start height above the sea level, length and riding velocity using the GPS, the requested crank power as well as the cardio respiratory adaptations (HR, oxygen consumption for energy expenditure calculation). The instrumentation used in order to record the data is: 1. dedicated handbike (Maddiline) with a power-meter (powertap g3 rear disc hub) set on the wheel hub; 2. metabolimeter (K4b2 Cosmed); 3. Heart rate, GPS, speed and altitude monitor (Garmin Edge 510). Participants' advices and clues to improve the track path (e.g. to change the slope of the track), the accessibility (e.g. more dedicated parking lots or more public transportation facilities) and the safety (e.g. reroute the pedestrian crossing) will be also collected. The analysis of the tracks has been performed on specific segment of it, to reduce variability due to transient phases in the beginning and in the end of the exercise.

4 DESCRIPTION

4.1 Innovative aspects

The research has been developed within an interdisciplinary team (MOTUS Project granted by University of Brescia). In particular, the results hereby showed, refer to the urbanistic and physiological outcomes which combine the measurements and the characteristics of the track paths with the ergometabolic engagement of a handbiker.

With regard to the former aspects a study for handcyclable track around the historical city centre of Brescia was performed as it follows: 1. elements of foreknowledge (standards for cycling tracks, handbikes data, handbikers experiences), 2. detailed analysis of the urban context (presence of green areas and their possible connections and geometric characteristics of the each single part of the track and their possible connections), 3. study of existing and planned cycle routes (mainly it consists of four different kinds of tracks most of which are not suitable for handbikes), 4. project of the track around the historical city centre (the proposal route cannot be used for sport purposes, but only for touristic and leisure ones because of the multiple constraints given by the historical context) (Arenghi et al. 2018).

4.2 Research and/or technical developments

Suitable handcyclable routes can be designed for a proper use outside the historical city centre where the existing context allows to have the geometric and services conditions above mentioned. This could be a good opportunity for Brescia municipality, while designing the new urban park in the south of the city, to improve accessibility for users with disability through an inclusive action.

5 RESULT OF THE RESEARCH

The graph in (Figure 2) reports the mean trends of some of the parameters recorded during the test by 10 subjects in a portion of a single lap of an exemplificative track. The trend of force and heart rate is strongly affected by the elevation profile. In (Figure 2) are highlighted two particular segments of the track: the main downhill (grey rectangle) and the subsequent uphill (yellow rectangle). As expected, the HR decreases following the descent and force reaches near zero values. Vice versa, during the ascent the HR rising is clearly noticeable. Force rising occurs just few moments later, this is probably linked to inertia previously developed. The fluctuations in the remaining sections of the track comply with these assumptions.

The third figure (Figure 3) shows the linear correlation between speed recorded by the GPS and the metabolic equivalent of task (MET) (1MET= $3.5mlO_2$/Kg/minute) calculated for every subject (blue spot) from direct measure of the oxygen expenditure using K4 device. The MET value refers to a single lap of the same track for every subject. It is clearly visible that the two variables (speed and MET) are directly correlated ($p<0.001$). Following WHO

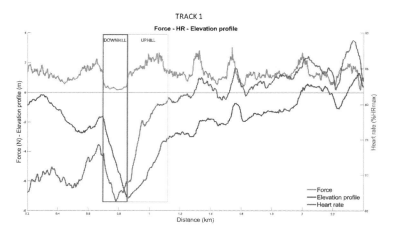

Figure 2. Elevation profile of the track and force applied to the cranks (left y axis) and heart rate (right y axis).

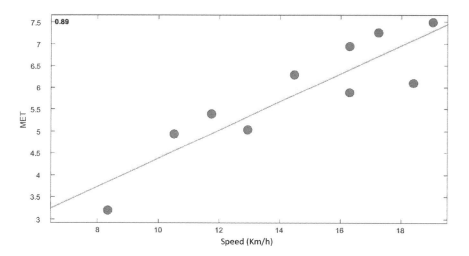

Figure 3. Correlation between speed and oxygen consumption standardised in MET.

recommendations (WHO 2010), to reach the correct amount of physical activity handbikers should cycle with an average speed between 7km/h and 15 km/h in order to perform moderate activity. Velocity over 15 km/h imply vigorous activity.

6 CONCLUSIONS

This project was aimed to sustain the local municipality to promote the practice of hand biking, in subjects with limited mobility, by targeted actions intended to make Brescia more accessible for adapted physical activity practice and to set a method for handbike tracks design and classification on the basis of the required effort. The tool described in this work could be considered for future institutional programs. The results of this pilot study in healthy subjects will provide a standard data-set to be compared with specific parameters obtained during hand-cycling in subjects with disability.

6.1 *Barriers and drivers*

In order to have handcyclable track paths, the Italian historical city centres are not suitable.

REFERENCES

Arenghi A., Piona M., Rosseti S., Tiboni M. (2018) Città e benessere: Pianificare e progettare lo spazio urbano secondo i principi di healthy city e active design in AA. VV. (2019), Atti della XXI Conferenza Nazionale SIU. CONFINI, MOVIMENTI, LUOGHI. Politiche e progetti per città e territori in transizione, Firenze 6-8 giugno 2018, Planum Publisher, Roma-Milano.

Barros RM, Silver EJ, Stein REK, et al (2017) The Association Between School-Based Physical Activity, Including Physical Education, and Academic Performance. J Sch Health 80:517–526. doi: 10.1542/peds.2007-2825

Donnelly JE, Ed D, Co-chair F, et al (2017) Physical activity, fitness, cognitive function, and academic achievement in children: A systematic review.

Ekelund U, Brown WJ, Steene-Johannessen J, et al (2018) Do the associations of sedentary behaviour with cardiovascular disease mortality and cancer mortality differ by physical activity level? A systematic review and harmonised meta-analysis of data from 850 060 participants. Br J Sports Med 1–9. doi: 10.1136/bjsports-2017-098963.

Rimmer JH, Riley B, Wang E, et al (2004) Physical activity participation among persons with disabilities: Barriers and facilitators. Am J Prev Med 26:419–425. doi: 10.1016/j.amepre.2004.02.002.

WHO (2010) Global recommendations on physical activity for health. World Health Organization.

Proactive city. the city as a gym for active design

E. Marchigiani, I. Garofolo & B. Chiarelli
Department of Engineering and Architecture, University of Trieste, Italy

ABSTRACT: The ageing of European urban population, the increase of social and economic inequalities, and the growing demand for supply and maintenance of public spaces and
welfare equipment foster deeper reflection on the material conditions supporting citizens'
health and mobility in the cities. Taking "accessibility for all" as a right of citizenship becomes
essential: the usability of urban environments plays a fundamental role in increasing
a person's ability to actively contribute to her/his own well-being. This perspective offers the
opportunity to reinterpret accessibility as a crucial component of people-centred and inclusive
urban design policies and tools. The on-going research PROACTIVE CITY outlines strategies
and solutions on these issues. Going beyond the simple removal or mitigation of the impacts
of physical barriers, the study focuses on the interpretation of cities as "gyms": places where
the choices related to the configuration of public spaces and the location of services are part
of strategies aimed at reactivating people's ability to move independently, to perform healthy
practices, and to interact.

1   A LONG LASTING DEBATE ON "ACCESSIBILITY FOR ALL"

Since the approval in 2001 of the WHO *International Classification of Functioning, Disability
and Health* (ICF), many studies and policies have contributed to raise the awareness of the
major role that spatial solutions play in making our cities accessible for the greatest extent of
users. The power that the interaction with urban environment has to enable people to manage
their own life has been mentioned by the *UN Convention on the Rights of Persons with Disabilities* (2006), and by the *European Disability Strategy 2010/2020* (2010). When promoting *Universal Design* (Bencini et al., 2018), these documents share the belief that disability is not
a condition proper to a person: it arises instead from the interaction between the individuals
and their everyday environment and, therefore, can temporarily or permanently affect everyone, in different phases of life.

Today, "Healthy Cities" (Tsouros, 2015; D'Onofrio, Trusiani 2017), "Active Cities"
(Edwards, Tsouros, 2008; AA.VV., 2015), "Inclusive Cities" (Shah et al., 2015), "Accessible
Cities for All" (Rossi, ed., 2018) are some of the locutions used for labelling national and
international city networks and experiences. Although the instances, actors and actions that
these practices refer to are diversified, they all foster reflection on the links between the configuration of city spaces and "mobility/motility" issues (Kaufmann, 2011; Sheller, 2019), and on
healthy living as a person's "capital" that is deeply influenced by the opportunity to autonomously and actively use urban places.

2   A PROACTIVE APPROACH

Although the need to integrate a variety of actions and disciplinary inputs is strong, many
barriers still divide specific competences, planning tools and urban policies dealing with accessibility: from the government of urban transformations and mobility, to the maintenance of
public spaces, to the provision of "territorialised" welfare services and equipment.

Since 2019, the interdisciplinary group Trieste Inclusion & Accessibility Lab (TRIAL), headed by the Department of Engineering and Architecture of the University of Trieste (DIA), has been focusing on the action research *PROACTIVE CITY. City as a Gym for Active Design*[1]. Going beyond the simple removal or mitigation of the impacts of physical barriers, PROACTIVE CITY focuses on the interpretation of cities as "gyms": places where the choices related to the configuration of public spaces and the location of services are part of strategies aimed at reactivating people's ability to move independently and to perform healthy practices.

Starting from some experiences developed in the Italian region Friuli Venezia Giulia, in this paper we will investigate both the convergences between current planning and design instruments and urban policies, and the obstacles that still oppose their more effective interaction. After describing the main research steps and its intermediate results, we will conclude with some addresses to rethink urban regeneration from the point of view of "accessibility for all".

3 EXPERIENCES AT THE BACKGROUND

At the background of PROACTIVE CITY lays a critical analysis of a number of experiences that, since some years, in our region have been focusing on accessibility as a driver for the innovation of public policies and tools.

3.1 *PEBAs and beyond*

In 2018, the Autonomous Region Friuli Venezia Giulia approved the law no. 1 on general principles and implementation provisions on accessibility, with the intent to review building and planning standards towards a safer and inclusive usability of public spaces and equipment. To support the implementation of the law, the Region activated: a technical consultancy center on accessibility; the construction of a geo-referenced accessibility mapping web portal; collaborations with the Universities of Udine and Trieste; funds to municipalities where a plan for the removal of architectural barriers (PEBA, Italian laws no. 41/1986, 104/1992) has already been approved. However, in the face of the limited amount of regional contributions (a maximum of 50,000 euros per municipality), this last condition is proving to be too restrictive, due both to the high commitment required for the elaboration of a PEBA, and to its recurrent focus on the removal of barriers to motor activity (while the solutions required by persons carrying perceptive and cognitive disabilities are often different and conflicting).

3.2 *Planning a framework of collective spaces*

In Friuli Venezia Giulia, the prevailing of small centers combines with the reduction – in line with national trends – of the size of households and the growing presence of elderly population. Consistently, in many recent town plans, the images of a "welfare of walking and everyday cycling" recur. The aim to make the urban structure more legible meets the intention of orienting public budget to the creation of a network of spaces open to the use of everyone. Starting from the large presence of equipment built as planning standards (green and sport areas, schools, social and health services, etc.), this physical framework works as a material support for the connection of public facilities, and for stronger synergies with design and sustainable mobility strategies (Basso, Marchigiani, being printed). What these experiences show is a promising trend towards the introduction of accessibility criteria and spatial solutions as components of the structural and regulatory contents of local town plans.

---

1. TRIAL is coordinated by Ilaria Garofolo. PROACTIVE CITY is a two-year research funded in 2019 by the University of Trieste, and sees the participation of: Elena Marchigiani (coordinator), Sara Basso, Barbara Chiarelli, Ilaria Garofolo, Lucia Parussini, Roberto Prandin, Valentino Pediroda.

3.3 Inclusive design

In 2011, the Province of Trieste established the Accessibility Laboratory (LabAc). Later integrated in the measures of the Plans for social and health care (Italian law no. 328/2000), LabAc involved various actors[2], with the aim of favouring inclusion and the autonomous mobility of the most fragile city users and inhabitants (Garofolo, Chiarelli, Grion, 2018; Garofolo, Marchigiani, 2019). On the side of the process, the planning of activities was based on the involvement of different skills, through the organization of surveys and of training courses (with professionals, public officers, researchers and persons carrying different disabilities). On the side of the outputs, LabAc led to a prototype of a digitalised and geo-referenced tool for mapping the degrees of accessibility of urban itineraries. In spite of the positive effects proved by the application of the LabAc's overall methodology, its transformation into ordinary practice has found opposition in the inertial nature of the sectorial and self-referenced routines that still prevail within the public administration.

3.4 Health paths

Percorso salute (Health path) is a participatory local welfare project, developed in the province of Trieste, and aimed at promoting physical activity in open spaces (Paoletti, Greco, eds., 2015). Started in 2016 by the Integrated University Health Service of Trieste (ASUITS), and involving local associations, the project organizes periodic meetings, where outdoor activities are performed together with health operators. In order to stress the importance to weave health and social policies, further initiatives have been recently developed through the organization of walking groups open to a wider public in social housing districts. Nonetheless, a problematic aspect is the lack of coordination with the public works dedicated to the maintenance and upgrade of the physical spaces where healthy activities take place.

4 AN ACTION (AND INTERACTION) RESEARCH

In order to help overcome the obstacles to "accessibility for all" that current tools and policies face, PROACTIVE CITY intends to integrate different research grounds (urban planning and inclusive design, rehabilitative physiotherapy, webgis and ICT, mHealth and e-care for neurodegenerative disorders, multi-criteria analysis), and develop strong interaction between reflection and practice, doing and learning. The first year of activity was dedicated to three parallel paths: the definition of addresses to planning and policy making; the drawing of spatial solutions for specific contexts; the reflection on technological devices for supporting decision on the priorities of interventions.

4.1 Guidelines

By taking part to INU national community "Accessible Cities for All", our research group contributed to the construction of a web portal, to be further implemented over time, containing *Guidelines for integrated accessibility policies and projects*[3]. The work was built on the analysis of more than 120 Italian case studies (the *Atlas*), focusing on planning/design/process solutions for the mitigation/removal of architectural, sensory, perceptual, cognitive, cultural, social, economic, health and gender barriers.

---

2. In addition to the University of Trieste with the role of scientific support and coordination (Ilaria Garofolo, with Elisabeth Antonaglia, Barbara Chiarelli and Silvia Grion), LabAc saw the participation of: Municipalities in the province; Territorial Agency for Social Housing (ATER); Integrated University Health Service of Trieste (ASUITS); Regional board of disabled people's associations.
3. See http://atlantecittaccessibili.inu.it.

Given the ineffectiveness of sectorial actions, and a general lack of extensive and medium-long term policies, the Guidelines give addresses to overcome current gaps in four operational fields: projects for a better usability of spaces and services; strategic and planning tools to foster accessibility; processes for urban policies' integration; training to promote awareness and interaction.

4.2 *Training and participatory workshops*

In line with an approach based on listening and testing, one of the core steps of the research was the identification of a pilot context in the region Friuli Venezia Giulia.

Grado was the place of a design workshop organized by DIA and the local Municipality, with the aim to integrate the issues of accessibility, walkability and cycling[4]. The choice was motivated by the complexity of a city that can be taken as representative of many other small and middle-size urban contexts in our region and country. Grado is a tourist destination, with a strong seasonal change of urban habits and population (from 8,000 inhabitants in winter, to 80,000 in summer). The local administration is implementing innovative projects for public spaces, with a specific attention to green infrastructures, sustainable mobility and resilience. The PEBA has been recently adopted, and the Plan for sustainable urban mobility is under construction.

During the two-week stay in Grado, professors and students met technicians from the Municipality, the Region Friuli Venezia Giulia, and took part into participatory surveys with the Regional board of disabled people's associations and the Italian Blind Persons' Union. Public training seminars on the issues of accessible cities were given by researchers and professionals involved in local and national planning and design experiences.

Meanwhile, students, professors and local technicians worked on design proposals for a new "green, healthy and accessible urban route", connecting the most residential part of the city to an existing system of parks and pedestrian areas at the back of the beach, where renewal interventions are needed. The objective was to conceive the itinerary as part of a broader and interconnected system of open spaces, paths for cycling and pedestrian mobility, areas for cultural, education and sports activities. Our challenge was to imagine Grado as a small capital of healthy and active life, offering the opportunity to perform different outdoor activities, all year long.

The workshop strengthened the awareness that "inclusive" is not synonymous with "special". The task is, therefore, to design without barriers from the very beginning, taking the perspective of the most fragile people to conceive spaces that are usable by everyone, and trying to solve the conflicts among different ways and capacities of movement (pedestrians *vs.* cyclists, motor *vs.* visual or cognitive abilities, as well as inhabitants *vs.* tourists).

4.3 *Platforms to share knowledge*

The progress towards digital revolution offers significant opportunities to innovate the structure of public organizations, the management of processes and of human and material resources. However, further work is needed to put into a system the many data produced and collected by the various sectors and levels of the public administration: i.e. the intensity of use and to the profiles of the users of public services, local public transport, accommodation facilities and sites of tourist-cultural interest; information related to the origin/destination of users, and to the demographic composition of residents in different urban neighborhoods (Marchigiani, Chiarelli, 2018).

In this direction goes the support that PROACTIVE CITY will give to the realization of the regional accessibility mapping web portal, as a tool offering useful knowledge for a multi-year programming of public works and services.

---

4. In July 2019, the workshop was coordinated by Elena Marchigiani, and tutored by Sara Basso, Barbara Chiarelli, Ilaria Garofolo and Valentina Crupi. Maria Antonietta Genovese was the reference person from the Municipality of Grado.

5 TOWARDS A DIFFERENT PERSPECTIVE

Research on innovative approaches and technical proposals involves opening up urban planning and design to a dialogue with many disciplinary contributions and actors. In this view, from now to its conclusion, PROACTIVE CITY will further investigate several action fields[5].

From people's disabilities, to new qualitative standards. The assumption of the "lens" of accessibility invites to rethink planning and building standards, in order to overcome their purely quantitative and functional conception, and to introduce new performance criteria to support the implementation of conditions for sustainable and inclusive mobility connecting public spaces and welfare services.

Coordinating general planning and spatial design solutions. Equally strong is the need to address the issue of accessibility within general planning tools, in order to build a reference vision for public works, housing, mobility and public transport programmes. This implies the review of administrations' routines, in order to break "silos thinking", and to enforce coordination between urban policies.

From designing for, to designing with. Design solutions should be ordinarily built through participatory processes, open to the contribution of all the "ignored experts" who everyday deal with the effects of disabling environments. In this sense, to make our cities "proactive", the research intends to enforce collaboration with actors involved in health interventions (i.e. physiotherapy), to better understand to what extent the removal of "all" obstacles is really effective, and what may constitute a stimulus to physical activity in relation to different pathologies and forms of disability.

Interacting and communicating with people, in a smarter way. ICT tools offer new opportunities to collect and share up-to-date information on accessibility conditions to urban spaces and facilities. The conception and testing of platforms that are able to return into spatial representations the many data that are already available to the public administrations can both address future interventions, and be used to communicate with citizens in a more customised and interactive way.

6 OPENINGS TO REFLECTION, BEING AWARE OF BOTTLENECKS

PROACTIVE CITY is still underway. Although we are aware that many topics are far too complex to be tackled by the research, we hope that showing the variety of bottlenecks in the relationships among actors, tools and processes can stimulate further study and experimentation on urban accessibility.

Specifically, developed experiences show the limits of binding innovative projects to the elaboration of PEBAs. A mapping of the physical obstacles to people's movement tends to reduce the complex theme of accessibility to a sectorial and remedial approach. On the contrary, his assumption as a structural component of ordinary urban planning and design can be much more effective.

Equally strategic is the coordination of timing and contents of general town plans with those of building regulations and of the many thematic programs that affect the usability of collective spaces (i.e. plans for traffic, mobility and parking, commercial occupation, etc.). At the same time, talking about healthy and active cities highlights the need to overcome the persistent misalignments between urban planning tools and social and health care plans ruling the territorialized organization of welfare services.

Finally, and as regards public works, accessibility projects can be implemented through time, and can therefore offer a way to cope with the current uncertainty of economic resources available for the regeneration of urban environments. Nonetheless, the conditions of their effectiveness are deeply rooted in the administrations' capacity to coordinate maintenance

---

5. In the frame of an agreement between DIA and the Autonomous Region Friuli Venezia Giulia, activities will coordinate with those of the board supporting the implementation of the regional law no. 1/2018.

operations, to clearly establish their criteria and priorities, to re-frame them into an integrated system of spaces and services.

All these considerations stress the need to renovate our technical skills Today, when referring to active and universal design, the discourse on accessibility often suffers from a contrasting tendency. On the one hand, the retreat into particular building and technological solutions confine the movement of persons with disabilities to dedicated spaces, thus producing a sort of "spatial stigmatization". In these solutions we can see the emerging of "new functionalisms" (Bianchetti, 2016): in the name of the application of ergonomic parameters that are "to the measure" of those who are more fragile, exceptional devices are used as additional ingredients to the design of open and built spaces. A second tendency is to a vague assumption of accessibility as a simplified attribute for connectivity, to be assigned to still traditional ways of designing open spaces and services.

What PROACTIVE CITY aims to show is that talking about "cities accessible for all" invites to rethink the design of urban spaces, tools and processes as places where people can return to play a central role, in virtue of their different bodies, motilities, material and immaterial needs, cultural and social habits and conditions.

REFERENCES

AA.VV. (2015). Active Cities. A Guide for City Leaders. Designed To Move, Nike inc.

Basso S., Marchigiani E. (being printed). Gli standard come capisaldi per ri-attrezzare centri urbani di piccole e medie dimensioni. Esperienze di pianificazione in Friuli Venezia Giulia. In: Territorio.

Bencini M.L., Garofolo I., Arenghi A. (2018). Implementing Universal Design and the ICF in Higher Education: Towards a Model That Achieves Quality Higher Education for All. In: Craddock G. et al., eds. Transforming Our World Through Design, Diversity and Education. Amsterdam, London, Washington: Diversity and Education. Amsterdam, London, Washington: IOS Press.

Bianchetti C. (2016). Spazi che contano. Il progetto urbanistico in epoca neo-liberale. Roma: Donzelli editore.

D'Onofrio R., Trusiani E. (2017). Città, salute e benessere. Nuovi percorsi per l'urbanistica. Milano: Franco Angeli.

Edwards P., Tosouros A.D. (2008). A Healthy City is an Active City: a Physical Activity Planning Guide. Copenaghen: World Health Organization.

Garofolo I., Chiarelli B., Grion S. (2018). Percorsi inclusivi e partecipati per la fruibilità degli spazi urbani: il caso studio LabAc. In: Angelucci F., ed., Smartness e Healthness per la transizione verso la resilienza. Orizzonti di ricerca interdisciplinare sulla città e il territorio. Milano: Franco Angeli. 307–324.

Garofolo I., Marchigiani E. (2019). Accessibility and the City. A Trieste, dispositivi e pratiche progettuali per attenuare le vulnerabilità sociali. In: Aa.VV. Atti della XXI Conferenza Nazionale SIU. Confini, movimenti, luoghi. Politiche e progetti per città e territori in transizione. Roma-Milano: Planum Publisher. 91–98.

Kaufmann V. (2011). Rethinking the City. Urban Dynamics and Motility. Lausanne: EPFL Press.

Marchigiani E., Chiarelli B. (2018). Con le 'lenti' della fruibilità: strumenti interattivi e tecnologici per rigenerare gli spazi urbani. In: Urbanistica Informazioni, no. 280-281. 96–98.

Paoletti F., Greco A.D., eds. (2015). Laboratorio di Welfare Locale Partecipativo. Trieste: Azienda Sanitaria Universitaria Integrata.

Rossi I., ed. (2018). Accessibilità, integrazione e scale: web, territori, città, quartieri. In: Urbanistica Informazioni, no. 280-281. 87–102.

Shah P. et al. (2015). World – Inclusive Cities Approach Paper. Washington D.C.: World Bank Group.

Sheller M. (2019). Mobility justice e le mobilità come bene comune. In: Perrone C., Paba G., eds. Confini, movimenti, luoghi. Roma: Donzelli editore. 45–58.

Tsouros A.D. (2015). Twenty-seven Years of the WHO European Healthy Cities Movement: a Sustainable Movement for Change and Innovation at the Local Level. In: Health Promotion International, vol. 30 (1). 13–17.

Pedestrians, Urban Spaces and Health – Tira, Pezzagno & Richiedei (eds)
© *2020 Taylor & Francis Group, London, ISBN 978-0-367-46171-3*

Elderly mobility under the microscope: A multidisciplinary perspective

E. Pantelaki, E. Maggi & D. Crotti
Department of Economics, University of Insubria, Varese, Italy

ABSTRACT: Mobility in later life is a key determinant in modern societies dealing with ageing population. Despite being studied in medicine, sociology and transport, this topic displayed findings which have been exploited within each discipline. Potential spill-overs have thus not been pointed out yet. Contributing to filling this gap, this study is a multidisciplinary systematic review aiming at informing about the impacts of elderly mobility on dimensions of well-being and quality of life. We searched for peer-reviewed articles published from 2010-2019 by scanning different electronic databases (Scopus, Web of Science, PubMed and TRID). 78 studies met the inclusion criteria, of which 70% come from medicine, 16% from sociology and 14% from transport literature. Interest-ingly, a large variability in ways to define the mobility of the elderly was detected, together with a number of discipline-specific measures and related effects. Since the topic of elderly mobility is indeed multidisciplinary fruitful sinergies among scholars are strongly encouraged in the future.

1 INTRODUCTION

Together with decreasing birth rates, advances in medicine and technology have pushed up life expectancy and are leading to ageing populations in both developed and developing countries (Cao and Zhang, 2016). In OECD countries, the share of people over 65 years old on the total population will reach 25.1% in 2050, from 7.7% in 1950 (OECD, 2015).

Nowadays, scientists and policymakers have shown growing attention on the issue of the ageing population, whereas in the past it was neglected. From an economic perspective, beyond raising concerns about the economic support for an increasing unproductive segment of the population, this concept mostly implies a variety of emerging implications on the health care (Abdullah et al., 2018; Aguiar and Macário, 2017), pension system, the general provision of consumer products and services in the society (Metz, 2000) and environmental issues (Aguiar and Macário, 2017).

World Health Organization supports the approach of healthy ageing which prioritizes the enhancement of functional ability by actively encouraging all relevant sectors to work together, including mobility policies (WHO, 2018; Musich et al., 2018).

2 OBJECTIVE OF THE PAPER

The aim of this paper is two-fold. On the one hand, to review systematically the literature on the effects of mobility on the elderly's life within different research fields (medicine, sociology and transport). On the other hand, to illustrate differences and similarities of their approach by bringing closer the results. In particular, this study aims to give answer to the following research question:

What are the effects of mobility on dimensions of well-being and quality of life (QoL) of the older people by scientists in medicine, sociology, and transport?

3   METHODOLOGICAL APPROACH

The systematic review was conducted by searching the relevant articles in the electronic databases: Scopus, Web of Science, PubMed and Transportation Research International Documentation (TRID). We searched for the keywords "mobility" together with the words "elder*","old*", "senior*","late* life" "age*" and "aging. We specifically selected these words because they are quite broad in order to retrieve articles deriving from the focused disciplines. The search started in May 2019 (titles only). The idea behind this method is that we aim at getting studies that are exclusively examining elderly mobility and not just referring to it at some points in the article.

The following criteria for inclusion were used: peer-reviewed studies written in English, community dwelling or non-institutionalized elderly people over 60 years old, living in developed countries and papers that examined any effect of mobility on older peoples' life. The methodology used is in line with the guidelines of PROSPERO[1], an international database of systematic reviews, and the followed protocol is awaiting registration confirmation.

4   DESCRIPTION

4.1   *Innovative aspects*

Mobility of the elderly people is widely studied by researchers and scientists in medicine, sociology and transport and there is consensus that mobility is crucial in later life. Although the researchers within the various disciplines uncover a diversity of effects for the elderly people, the findings are mainly exploited in an intradisciplinary way. To the best of our knowledge, this is the first multidisciplinary study to stress the effects of mobility on the life of the elderly people. The advantage of a multidisciplinary approach is that it could reveal aspects of the topic that otherwise remain hidden (Murray, 2015), since the perspective of each discipline is substantially different.

4.2   *Research*

Searching in the electronic libraries, 3416 English articles were retrieved. After applying the set criteria, 78 articles (55 medicine, 12 sociology and 11 transport) were considered as relevant to answer the research question.

The research has highlighted that the toolkit of mobility measures is quite broad and, more specifically, there are significant differences in defining and measuring elderly mobility by researchers not only between but also within the disciplines. Given that there is not homogeneity of the mobility measures, this leaves room for high sensitivity of the investigation of the effects of mobility on aging well-being and QoL.

5   RESULT OF THE RESEACH

Related findings are presented for each discipline, as follows.

MEDICINE: Medical researchers are interested on the impact of mobility as a physical ability for the performance of daily tasks and how it affects dimensions of well-being and QoL. The toolkit for mobility measurement is rich and contains both objective tests of functional ability and subjective self-assessments (through answers to targeted questions). The aim of the studies is basically to explore impacts on health conditions and to a lesser extent maintenance of independence and enhancement of well-being and QoL.

---

1. https://www.crd.york.ac.uk/prospero/

SOCIOLOGY: The sociological literature shows a mixed panorama. Mobility has a broader meaning of movement on the physical space, considering the means used to move or difficulties, and impairments in movement. The systematic review shows that mobility is found to be beneficial for health, independent living and social inclusion in later life.

TRANSPORT: In transport research, mobility is connected to travel modes that facilitate the movement and measurement tools were detected accordingly. Although health, social inclusion and QoL implications were found, substantial and further empirical research is essential to clearly frame the role of mobility in later life.

6 CONCLUSIONS

The ageing of the worldwide population indicates that targeted research is crucial to give directions to policymakers on how promote effectively healthy ageing. Elderly mobility is a multidisciplinary research topic and should be treated in that way by researchers. It is a promising research field and our approach points to this direction.

The analysis outlines that there are some research gaps that can be afforded by further research efforts. First of all, there is the need to connect the effects of mobility with related measures. Since elderly mobility is measured by different ways, a comparison of two or more tools with respect to the same effect may tell us whether the measure selection is sensitive to the results. By this study, it is possible to create a multidisciplinary measurement for elderly mobility. Finally, for each type of mobility it could be explored its effect on health, independence, social inclusion, well-being and QoL (Active Ageing Index, 2013).

6.1 *Barriers and drivers*

Our systematic review has some strengths and some limitations. As regards the first ones, to the best of our knowledge, this is the first multidisciplinary approach to the topic of elderly mobility. Presenting in the same time the results of the three disciplines is a strong motivation for future fruitful collaboration.

Concerning the limitations, one of the drawbacks of the systematic review is that it presents only the state-of-the-art literature. However, there might be relevant studies that were published before 2010. To the authors' knowledge, since it is the first multidisciplinary review an earlier period multidisciplinary approach will be able to present the evolution of the topic. A second limitation is the selection of the keywords and the searching methodology. Although the selection was on purpose generic, so that we can get as much as possible studies, probably a different keyword selection could return back slightly different results. Finally, the search of the key terms considered only the article titles because of our intention to get studies that were entirely devoted to the impacts of mobility. However, there might exist studies appropriate for inclusion that do not contain the combination of our selected keywords.

ACKNOWLEDGEMENT

This review is the first step of the research project HAPPY, funded partly by Cariplo Bank Foundation (id number: 2018-0829) and partly by the University of Insubria (PhD scholarship for Ms Pantelaki).

REFERENCES

Abdullah, N.N., Ahmad Saman, M.S., Kahn, S.M., Al-Kubaisy, W.,(2018).Older People with Mobility Disability (Quality Of Life). Asian Journal of Quality of Life 3. 103. https://doi.org/10.21834/ajqol.v3i11 126.

Aguiar, B., Macário, R.,(2017). The need for an elderly centred mobility policy. Transportation Research Procedia 25. 4355–4369. https://doi.org/10.1016/j.trpro.2017.05.309

Cao, J., Zhang, J.,(2016). Built environment, mobility, and quality of life. Travel Behaviour and Society 5. 1–4. https://doi.org/10.1016/j.tbs .2015..12.001.

Metz, D..,(2000). Mobility of older people and their quality of life. Transport Policy 7, 149–152. https://doi.org/10.1016/S0967-070X (00)00004-4.

Murray, L., (2015). Age-friendly mobilities: A transdisciplinary and intergenerational perspective. Journal of Transport & Health 2, 302–307. https://doi.org/10.1016/j.jth.2015.02.004

Musich, S., Wang, S.S., Ruiz, J., Hawkins, K., Wicker, E., (2018).The impact of mobility limitations on health outcomes among older adults. Geriatric Nursing 39. 162–169. https://doi.org/10.1016/j.gerinurse. 2017.08.002.

OECD, (2015). Ageing in cities. OECD. Paris.

World Health Organization, (2018). The Global Network for Age-friendly Cities and Communities: Looking back over the last decade, looking forward to the next.

Participatory experiences supporting more healthy and active cities. the research-intervention "Anziani&Città"

E. Dorato
Department of Architecture, CITERlab, University of Ferrara, Italy

ABSTRACT: This contribution investigates the complex and multifaceted relationships and factors of influence between the aging process of the so-called "urban seniors" and the physical and perceptive characteristics of the city. By developing an ecological model of aging based on the evidence of interdisciplinary scientific research, Urbanism is intended as a fundamental preventive discipline capable of greatly improving people's quality of life, while contributing to the maintenance of health, sociality and overall autonomy. Supporting this theoretical framework, the outcomes of the qualitative research-intervention *Anziani&Città* are presented: a work by CITERlab financed in 2018 by the Emilia-Romagna region with the aim of developing and testing a three-phase participatory process with groups of senior citizens in different cities and neighborhoods, investigating the multiple relations between the urban built environment and the elderly's quality of life: their health, practiced levels of health-enhancing physical activity and active mobility, perception of safety and urban accessibility.

1 THE ELDERLY AND THE CITY: CONTEMPORARY ISSUES AND CHALLENGES

Urban planning and design are disciplines with the great power to influence human behaviors and quality of life. Evidence has and still is broadly demonstrating that the ways in which cities are designed and structured can have both a physical and mental impact on the people who live, work, play, love, and grow old in them. Given the unprecedented demographic transition we are globally witnessing, the entering of the baby-boomer generation into the old age represents not only an important health care, social, political and economic dare, but also and especially a great *urban* challenge. A challenge for our cities, for the effects that the consolidated urban models – in terms of livability, connectivity and transport systems, delocalization, etc. – have and will have on the health and wellbeing of urban population; for the direct effects that aging, migratory dynamics, new urban social compositions, and the evolution of the concept of fragile population will have on the design and use of public and collective spaces, with repercussions on public health; for the economic burden associated with severe demographic imbalances; and for the reduced financial resources that will limit public investment and the possibilities for good project implementation at the urban scale. In recent decades, research has focused on the elderly's living conditions, namely the almost "curative" technological and home automation aspects of the building envelope and its components, yet neglecting the collective and urban dimension. However, the need for more accessible, safe, flexible and inclusive urban spaces, responding to a growing diversification of needs and social behavior, arises from civil society.

It is a proven fact that public, collective and connective spaces and their characteristics play a fundamental role in maintaining the physical, intellectual and social autonomy of the elderly, as well influencing the entire population life quality. Also, scientific research tells us that a sedentary lifestyle in senior population, in addition to favoring the onset of chronic, neurodegenerative and musculoskeletal disorders and disabilities (WHO, 2002; 2007), is exacerbated by psychological dynamics linked to physical and social isolation. It is necessary to reverse the consolidated and prevailing logic of spatial and social organization of senior population "founded either on the adaptation of the elderly to the existing, or on their exclusion" (Amendola, 2011), modifying and adapting the urban habitat to the peculiar and fast-changing needs of senior citizens. To promote active aging and facilitate the daily experience of the city, the urban spaces themselves must adapt and transform, responding to policies and programs aimed at detecting weaknesses in the existing urban system and, consequently, improve it. Therefore, planning and designing for accommodating the widest possible range of elderly needs could bring great benefits to the entire population, for preventing the arise of vulnerabilities represents the most effective and economic option. In this perspective, the Active City model focuses attention on the physical, spatial and socio-cultural re-appropriation of public spaces by the people (Dorato, Borgogni, 2018; Borgogni, Farinella, 2017), as well as on the preventive role of Urbanism, shifting the current and widespread "medicalized" approach (Borasi, Zardini, 2012) to a "socio-ecological" one, in the belief that an active city is also a healthier, safer, accessible, pleasant, sustainable and inclusive city. This means thinking or re-thinking urban space as a construct capable of enabling any citizen, especially if over65 and with dynamically changing needs, putting in place new tools aimed at meeting the required performances rather than prescriptive ones (Ghirardelli, 1996).

2   MAIN OBJECTIVES OF THE RESEARCH-INTERVENTION

Consistent with a series of works carried out since 2015 by *CITERlab* (Architecture Department, University of Ferrara) for the Urban Quality and Housing Policies Service of the Emilia-Romagna region, the research-intervention *Anziani&Città* had the main objective to qualitatively investigate the habits and needs of a sample of elderly population in terms of life quality; perception, use and livability of public spaces; health conditions; and physical activity practice, in relation to the use and characteristics of urban public spaces, focusing on the pilot cases of Reggio Emilia and Bologna. Working together with two heterogeneous "convenience groups" of self-sufficient 65+ citizens, the research has experimented with a three-stage participatory process, potentially replicable by methodology and purpose in other cities. Through the active involvement of elderly people, using mixed techniques and tools such as semi-structured qualitative questionnaires, focus groups and participatory planning sessions, *Anziani&Città* aimed at formulating proposals for small urban transformations, better responding to the real needs of seniors, believing that a city conceived and re-designed *with* and *for* the elderly can be – in a global perspective of sustainable urban development – a city for all. Therefore, the identification of local needs deriving from the layout of roads, traffic and public and active mobility (cycle and pedestrian); the main weaknesses in the connective system and urban public spaces accessibility; the habits in the practice of physical activity (not only as a recreational and social moment, but also as a key factor for a healthier aging) represent the main challenges as well as the focal points of the work. It has no statistical purposes, as numerical data collected from questionnaires and qualitative information deriving from the focus groups served only to better understand the habits and needs of the samples, guiding the meta-design phase. This work represents a perfectible process of investigation and participatory planning which could be enriched, during future implementations, with new inputs aimed at optimizing results and improving the whole process.

3   RESEARCH APPROACH AND STRUCTURE

Thanks to preliminary consultations with local stakeholder[1], the two cities and related pilot neighborhoods of Bologna San Donato and Reggio Emilia Ospizio were identified according

---

1. The Emilia-Romagna region and the Social Policies/Urban Planning Services of the two municipalities were involved, together with local public institutions managing social services (ASP), whose early involvement was crucial for the *effectiveness and success of the whole process.*

to some main criteria, such as the peripheral location of the districts; a strong presence of elderly resident population; and the pre-existence of organized groups for socio-recreational or physical activities in public-run facilities and spaces. The senior citizens regularly taking part in the research-intervention were 14 in Reggio Emilia and 10 in Bologna; they freely chose to participate, informed by local operators or by friends and neighbors, which allowed the two self-selected groups to be particularly sensitive and interested in the specific topics dealt by the research. The work has been structured in three main phases, each of which corresponding to a meeting with the convenience groups to answer specific questions and needs. In order to draw a picture of the possible impacts of the built environment and its characteristics on urban quality, the livability of public spaces, people's perception of health and practiced levels of physical activity, during the first meeting all participating subjects were asked to complete an semi-structured qualitative questionnaire. This tool, composed of 9 open-ended questions and 34 closed-ended questions, was created *ad hoc* for the study, using parts of validated European survey tools[2], with the addition of some introductory questions and some final questions to further detail perception and appreciation of the area of residence, in line with the characteristics of incidence of the urban environment already highlighted by scientific evidence. Deepening the most relevant issues emerged from the questionnaires, the second phase focused on the detection of needs, through the direction of focus groups. All participants actively took part to the discussion, explaining and detailing their point of view regarding specific problems in their neighborhoods: real and perceived urban safety; accessibility of public and collective spaces; vehicular traffic, public transport, noise and environmental pollution; impairments connected to travel, the practice of physical activity, and the performance of daily activities in the urban context; social and relational difficulties. After collecting a considerable amount of information and points of view during the first two phases, the last phase used images and cartographic bases at different scales (urban, neighborhood, micro-area) for intervening with drawings, colors, symbols and post-it containing opinions and meta-design proposals. It is important to underline how, although no economic budget was established for possible future urban transformations, both groups autonomously focused on low economic impact proposals, increasing the feasibility of urban adaptation works.

4 RESEARCH DATA AND MAIN OUTPUTS

Consistently with Giuliani's theory (1991; 1995), sustaining that "place attachment" in the elderly derives both from the meaning places have for the subject's personal identity, and from the capacity of that particular place to satisfy one's needs, the degree of satisfaction with respect to the urban environment of residence has been far greater in people who have resided longer in the reference neighborhood, while negative judgments were more evident in those who have lived in the neighborhood for less than two years. As Iecovich (2014) discussed, seniors' perception often appraises the judgment on functionality and adequacy, loading the term *place* with multiple and closely interconnected meanings. Women represent the vast majority of the sample and more than 70% of the participants live alone, with an average group age of 79.5 in Bologna and 77 in Reggio Emilia. In a nutshell, the general conditions of roads and buildings (aesthetics, degradation, cleanliness, presence of trees, etc.) were not considered adequate in either contexts, as well as the perception of air quality and noise pollution. Generally inadequate or poorly maintained also the conditions of sidewalks and cycle paths; the urban elements in public spaces are deemed to be poorly suited to the needs of the elderly; cycle-pedestrian intersections are perceived as unsafe and the duration of pedestrian crossing green lights decidedly

---

2. Special Eurobarometer 378 "Active Ageing" (QB2, QB1, QB29, QB30); Flash Eurobarometer 419 "Quality of Life in European Cities" 2016 (Q1, Q1.1, Q1.2, Q1.5, Q1.6, Q1.7, Q2.8, Q1.8, Q1.10, Q1.11, Q1.12); Special Eurobarometer 412 "Sport and Physical Activity" (QD1, QD2, QD4b, QD5a, QD5b, QD7, QD9, QD9, QD11.1).

insufficient. In Bologna San Donato actions to increase the quantity and quality of urban public spaces are needed, and the perception of safety within the public spaces is low; however, the urban public transport system is judged very positively, unlike Reggio Emilia Ospizio where buses and related stops were not considered accessible and safe. Here, the perception of safety in the neighborhood is higher, while local retail (variety of offer, location and accessibility of stores) represents a critical issue in performing daily activities independently. In both contexts, the conditions highlighted as crucial for making neighborhoods more age-friendly were *"more facilities where senior citizens can keep active and healthy"*, and *"interventions on roads to increase road* safety". Supporting the very aim of the research, the question on where physical activity is usually performed highlighted a prevalent answer as *"in the park, in the garden or in the open air"*, followed by *"during daily journeys home-places of interest/for running errands and* purchases". Such qualitative information confirms that urban public spaces represent, especially for the elderly, the privileged context where to carry out physical activity, as well as the scenario for maintaining their social relations. In terms of urban design proposals, both groups decided to focus mainly on traffic and road-related issues, by proposing transformations on certain dangerous intersections, better regulating the different flows; and on the re-configuration of specific public areas which, at present, do not provide the adequate and most needed features.

5   REFLECTIONS AND FUTURE PERSPECTIVES

Virtuous experimentations and theoretical and applied studies are showing how the dyad"health and welfare system - housing policies" today must necessarily be integrated by a thirdand equally relevant element: the system of urban public spaces. These are the three pillars onwhich to ground and build policies and interventions supporting active aging and, more generally,the elderly population, while also contrasting the decline and degradation of many urbanareas (Martinoni, Sassi, 2011). Evidence is consolidating around the many works and researchesat European level that are questioning the criteria and strategies on which to base interventionsand projects of urban adaptation and/or transformation aimed, primarily, at the creation of agefriendlycities and neighborhoods, producing a multitude of documents and guidelines. Thanksto such suggestions, the discipline is trying to find solutions to complex issues on approaches,processes, and projects quality and characteristics aiming at improving urban livability and, consequently,simplify the city daily experience of older people, focusing increasingly on the peculiaritiesof public space as a determinant of health and active aging. The consensus is growing onthe impacts that the urban built environment has on the mobility, independence and quality oflife of the elderly, with effects also on the possibility of aging as autonomously as possible withinone's own physical and social context. Today, international literature almost unanimouslyagrees on three macro-conditions considered fundamental for a better quality of life of the elderlyin cities, factors found and widely discussed also during the research-intervention Anziani&-Città: widespread accessibility; perception of safety; and a well distributed presence of spacesand services that allow and encourage open-air movement and socialization (Figure 1).

Figure 1.   Areas of intervention and most relevant criteria in the definition of age-friendly urban contexts, as emerged from the first experimentation of the research-intervention *Aziani&Città.*

Starting from the contextualization of these three factors, from the study and application of scientific literature results and data, and from the outcomes of the participative process carried out in the pilot cities of Bologna and Reggio Emilia, it was possible to define ten main reference criteria, potentially integrable on the basis of the results of further experimentations in other cities. Such criteria could provide a solid and scientifically updated basis for intervening in urban contexts, making them more adequate and responsive to the needs of the elderly population and, at the same time, more pleasant, safe, accessible, healthy and inclusive for all. However, despite the fundamental importance of advancing in this research fields both quantitatively and qualitatively, political will still represents the greatest asset and, at the same time, the principal obstacle to a broader and more effective conceptualization of the *Active City* and its implementations focused on the elderly (Borgogni et al., 2017). The main barrier is still represented, both at political and academic level, by the lack of cross-disciplinary policies, actions and researches.

REFERENCES

Amendola, G. (2011). Abitare e Vivere la Città. In: Golini A., Rosina A. (a cura di), Il Secolo degli Anziani. Come cambierà l'Italia. Bologna: Il Mulino, 97–111.

Borasi, G., Zardini, M. (2012). Imperfect Health: The Medicalization of Architecture. Zurich: Lars Muller Publishers & Canadian Centre for Architecture.

Borgogni, A., Farinella, R. (2017). Le Città Attive. Percorsi pubblici nel corpo urbano. Milan: Franco Angeli.

Borgogni, A., Dorato, E., Arduini, M., Ciccarelli, M., Digennaro, S., Farinella, R. (2017). Active and sociable cities: the frailer elderly's perspective. In: Caudo, G., Hetman, J., Metta, A. (eds.), Compresenze. Corpi, azioni e spazi ibridi nella città contemporanea. Rome: Roma TrE-Press, 45–47.

Dorato, E., Borgogni, A. (2018). The Active City perspective. A.C.C. as an active planning tool. In: Pezzagno, M., Tira, M. (eds.), Town and Infrastructure Planning for Safety and Urban Quality. London: Taylor & Francis Group, 53–60.

Iecovich, E. (2014). Aging in Place: from Theory to Practice. Anthropolocigal Notebook, 20(1), 21–33.

Ghirardelli, M. (1996). La Città e gli Anziani: una lettura critica delle normative italiane in materia di accessibilità urbana. Paesaggio Urbano, 2/1996, 60–63.

Giuliani, M.V. (1991). Toward an analysis of mental representation of attachment to home. The Journal of Architectural and Planning Research, 8, 133–146.

Giuliani, M.V. (1995). Il ricordo dei luoghi nella memoria autobiografica. Ricerche di Psicologia, 19(2), 35–49.

Martinoni, M., Sassi, E. (2011). Progettare uno Spazio Pubblico a Misura di Anziano. In: AA.VV. Access SOS. Costruire città accessibili a tutte le età. Strumenti e azioni. Ferrara: Corbo Editore.

World Health Organization (2002). Active Ageing: a policy framework. Geneva: WHO.

Promoting healthy cities

Pedestrians, Urban Spaces and Health – Tira, Pezzagno & Richiedei (eds)
© 2020 Taylor & Francis Group, London, ISBN 978-0-367-46171-3

The missing path: Promiscuous bicycle lanes in urban areas

R. Fistola & M. Gallo
Department of Engineering (DING), University of Sannio, Italy

R.A. La Rocca
Department of Civil, Architectural and Environmental Engineering, University of Naples Federico II, Italy

R. Battarra
Institute for Studies on the Mediterranean, National Research Council, Naples, Italy

ABSTRACT: This paper deals with urban cycling considered as a factor of improvement of urban sustainability, if adequately supported by appropriate and efficient infrastructures. Starting from a classification of some conditions meant as the main dyscrasias that can occur analyzing a generic urban bikeway, this study proposes a methodological procedure aimed at formalizing a measure of the efficiency of the urban infrastructure dedicated to the soft mobility. The procedure pointed out proposes a measure of dyscrasias by considering not only the length in kilometers of urban bikeways, but it tries to indicate a set of variable able to express the real usability of the cycle paths. The procedure, here in its first formulation, could be considered as a tool to verify the condition that can improve the use of the bicycle as means of transport in an urban context, enhancing the urban livability through the optimization of the sustainable mobility.

1 INTRODUCTION

The theme of cycling in urban areas is still a challenge for the Italian cities (Mantuano et al., 2017) that despite their popularity for being part of a precious cultural and historical heritage, have bad performances as green and sustainable cities. They are also famous for being related to the production of urban pollution and vehicular congestion (Fistola et al., 2012). Some of these negative records are annually highlighted in several reports as the Legambiente *Urban Ecosystem* (2018) that depicts a well-structured picture of the Italian cities, posing particular attention to the environmental indicator for the urban quality of life. The section of the report dedicated to urban mobility shows that Italy has the highest density of cars (37 million) within the European countries and it is stressed that the congestion of the mobility networks cost 142 billion euros of GDP over the last ten years. The most significant damage is produced by urban, local and commuter mobility, which absorbs 97% of all journeys. To offer new possibilities of soft mobility (also thanks to various forms of funding), several Italian cities are providing themselves with urban cycle paths as an alternative to private cars. Numerous studies have shown that investments in cycle roads are amortized over a sufficiently short period and that they can produce positive effects for the city. Some other studies (Isfort, 2016), showed that there is a strong propensity of urban populations to renounce to the private vehicles and to utilize alternative ways of moving (Isfort, 2016). Nevertheless, many Italian cities have not yet approved an Urban Sustainable Mobility Plan (PUMS) as prescribed by the new European guidelines and by the decree-law in August 2017. PUMS, indeed, could represent the turning point both to modify the urban layout and to integrate the urban planning targets with the mobility planning, as well as to be a basic condition for allocating the European funds intended to support the promotion of the sustainable mobility.

2 OBJECTIVES OF THE PAPER

The main objective of this study is the attempt to develop an indicator capable of synthetically expressing the real effectiveness of cycle paths in urban areas. This stems from the need to investigate a circumstance that characterizes Italy: i.e. the increase in cycle paths do not lead to an increase in urban cycling by default. In Italy, in fact, despite an increase of almost 50% of cycle paths in the last years, the use of the bicycle as a modality of travel in cities is still not widespread. This limited interest maybe is due to some structural deficiencies of the routes that also generate in the users the perception of low safety. It could be stated that to be considered as affordable and comfortable, the cycle path must have continuity and connective qualities (Cole-Hunter et al., 2015). Often these qualities are thwarted by several anomalies that can be individuated and then analyzed in order to set up a taxonomy of "irregular cases" that allows for the working out of a method to assess the real effectiveness of urban cycle routes. In this regard, the present study aims to introduce a scientific method based on the analysis of the physical and functional characteristics of an urban bikeway network and to compare these characteristics with a set of dyscrasias corresponding to the conditions that could prevent the real use of the cycle path. By the assessment of the "dyscrasias" that can arise along the path, it is possible to pinpoint adapt interventions to optimize the use of urban cycle paths (Muñoz et al., 2016).

3 ASSESSING THE "MISSING" PATH

In order to assess the real effectiveness of cycle paths in urban areas, a set of conditions that hinder the use of the cycle path was identified. These conditions have been defined as anomalies that can be detected and weighed to express an "indicator of discrepancy" of the use of the path. Based on the analysis of many examples of exiting urban bikeways (Figure 1), a set of 10 dyscrasias/anomalies have been categorized in the followings:

DYSC. 1   poor interconnection between the path and the main mobility poles within the urban context (the path does not connect each other the points of the urban mobility network like stations, parking, etc.);

DYSC. 2   width change of the path (path section size);

DYSC. 3   uneven surface of paving (not-smooth paving path);

DYSC. 4   poor maintenance of the path and lack of ancillary services;

DYSC. 5   ghost tracks (sections of the track are not connected);

DYSC. 6   disruption of the path (the path has not a starting or ending point, but it suddenly stops);

DYSC. 7   promiscuous or dangerous crossing;

DYSC. 8   unsuitable promiscuous section (dangerous co-presence);

DYSC. 9   presence of obstacles due to the misuse of the path (parking on the track, anomalous use of the path, e.g. wheelchairs, strollers, etc.);

DYSC. 10 lack or poor signage and services (horizontal, vertical, lighting, acoustic).

Figure 1. Two examples of dyscrasias. On the left a bike path with promiscuous use, unprotected from the adjacent carriageway, which deviates sharply towards a dangerous road-crossing poorly flagged. On the right a misuse of the path.

Detecting the anomalies listed throughout an urban cycle path, it would be possible to set up an algorithm able to provide a synthetic expression of its inefficiency. The detected anomalies, considered with different weights, could be related to the territorial parameters (resident population, urbanized surface, length in km of the urban road network, etc.) according to the availability of data that describe the urban context.

4 DETECTING ANOMALIES: THE CASE OF NAPLES

In this part, a first attempt to test the methodological considerations has been developed. The study considered the case of Naples as a meaningful sample both for the check of the anomalies that can be observed, and to validate its compliance with the image of a smart and sustainable city. The exiting bikeway in Naples has been analyzed referring to the metropolitan municipality ordinance concerning the sustainable mobility (2012) that declares the institution of a cycling mobility path connecting the city from Bagnoli (west side) to Piazza Garibaldi (east side)for a total length of 20 kilometers. The main characteristics of the path are illustrated in Table 1 according to the description of the document (Comune di Napoli, 2012), while the last two columns refer to Legambiente[1] criteria, assigned to the cycleway of Naples.

In a second phase, a direct analysis was developed to verify the conformity between the described path (document) and the real condition of its usability, in its present configuration. The data coming from the field investigation have been compared with the ten typologies of dyscrasias before mentioned, in order to individuate the weakness of the path, that indicate the points or the sections in which there is a condition of priority of intervention. Table 2 shows the relation between the composition of the cycle path and the dyscrasia occurrence calculated in percentage of the total length of the path.

In its present configuration, the urban cycle path in Naples presents dyscrasias mainly connected to the lack of adequate signage. 35% of the path is not well marked especially in mixed section (bicycle/pedestrian) characterized by a huge density of urban activities (e.g. along the commercial area of via Toledo and via Chiaia). 15% of the exiting path is characterized by dangerous intersections between vehicular traffic and cycling; in fact, some sections must be travelled carrying the cycle by hand (e.g. closed to Galleria Laziale).

Table 1. Table shows the characteristics, per sections, of the cycle path in Naples.

Composition of the path	N° of sections	% of tot km	Total length of the sections	Multiplier by Legambiente	Total points
Mixed pedestrian-bicycle sections	16	15%	3km	0.3	4.8
Sections obtained from the sidewalk	4	30%	6km	0.2	0.8
Sections in reserved lane on the vehicular streets	6	10%	2km	0.5	3
Cycle crossing	10	10%	2km	0.3	3
Mixed cycle/pedestrian crossing	3	3%	0.6km	0.3	0.9
Sections where get off the bike and carry it by hand	2	1%	0.2km	0.3	0.6
Reserved lanes	3	30%	6km	0.2	0.6
Zone 30	2	1%	0.2km	0.2	0.4

---

1. In the report Urban Ecosystem elaborated by Legambiente different sections composing a cycle path in urban context are classified as follow: 0,5 points if the path has a reserved lane; 0,3 points if the path has mixed sections (bicycle/pedestrian); 0,2 points if the path is within a Zone 30.

Table 2. Percentage of the path composition where dyscrasias occur according to the typologies defined before.

(% of tot lenght)	DYSC 1	DYSC 2	DYSC 3	DYSC 4	DYSC 5	DYSC 6	DYSC 7	DYSC 8	DYSC 9	DYSC 10
Naples bikeway	0%	7%	15%	1%	3%	5%	15%	10%	9%	35%

Sections that are along the vehicular streets have not yet their reserved lanes worsening their promiscuity with the car traffic. At the current state, the results of the field test have not yet considered the weight that dyscrasias can assume, therefore, some considerations can be elaborated based on the analysis of the characteristics of the path.

Dyscrasias related to the wrong design of the path can be clustered as *incoherent* as an urban path has to relate to the main pole of the urban mobility in order to propose itself as an alternative infrastructure to improve sustainable urban mobility.

Other dyscrasias linked to the dangers that can occur, especially in the mixed section, can be clustered as a condition that transmits a *lack of safety* for users. Dyscrasias connected with the sudden interruption of the paths highlight the absence of a coherent design at the basis of the concept of the path.

5 TOWARDS A METHODOLOGICAL FORMULATION: AN INITIAL PROPOSAL

The anomalies individuated before can be related to the definition of typological clusters that correspond to the level of inhibitions for the use of the cycleway. In this regard, an initial list of a typological cluster can refer to:

- anomalies depending on design, i.e. absence of curbs, etc.;
- organization anomalies, i.e. missing or poor signage, etc.;
- anomalies of intersections, i.e. the excessive number of crossings; etc.;
- obstacle anomalies, i.e. incorrect use of the path as parking or other.

The first cluster of anomalies has higher weight than others, according to the objectives of this study. In this regards, a significant indicator of measuring could be the equivalent kilometers, but it has been considered that the exclusive measurement of kilometers is not sufficient to assess the impacts of these anomalies on urban livability, as the quality of the paths is strongly connected to its design, rather than to its length solely. For the assumptions of the present study, an ideal[2] urban cycleway should meet two main kinds of requirements:

1. the design of each segment making the cycle path;
2. the care of the connection between the segments that make the cycleway and between these and the network of streets.

The first requirement concerns the conception of the path and of the single segments that will constitute it, having care of assuring high standards of design and safety, with particular reference to the width of the lane, the level of separation with other flows (cars, buses, pedestrians, etc.), the type of pavement, lighting and protection of crossings (traffic lights), signals and availability of parking spaces. The dyscrasias, or irregularities, could lead the cycle path far from these "ideal conditions".

---

2. In this context, ideal refers to optimal condition of the path (good design, absence of point of danger, lack of irregular intersections, adequate use, etc.).

The second point concerns the connections between the segments and among paths. For instance, it is very hard to cycle on a path made by several segments not well connected each other, because to switch from one segment to another it could be necessary to travel on not-reserved routes. While to travel on a path that is included in a well-connected network can be very easer as well as safety and comfort. Therefore, in this second case, the path is entirely for cycling assuring to never leaving the protected lane.

In the context of these considerations and even though in a preliminary state, it has been proposed a methodology able to measure the real effectiveness of a bicycle path, considering an "equivalent length". In practice, considering 10 km of fully connected cycle path having high standards of design and safety, its equivalent length is 10 km, that is the real and equivalent length coincide. On the other hand, if the path diverges from the optimal conditions, the equivalent length will be lesser than the real length. In general, the equivalent length of the path could be calculated with the following equation:

$$L_{eq} = \sum_j (L_j \cdot k_1 \cdot k_2 \cdot \ldots \cdot k_n) \cdot C_{con}$$

where:

L_{eq} is the equivalent length of the path (km);

j indicates the generic segment of the path with homogeneous characteristics;

L_j is the length of the homogeneous segment j;

$k_1, k_2, \ldots, k_n$ are coefficients less than 1 that measure the distance between the segment features and the ideal ones (these coefficients are equal to 1 for the ideal case);

C_{con} is a coefficient less than 1 that measures the distance of the path configuration concerning an ideal case of a fully connected network.

In the further development of the research, methodologies will be studied for assigning values to correction coefficients. It should also be pointed out that the procedure can be improved by adding specific assessments of the usefulness of the planned connections, depending on the mobility demand. In this phase, some general indications on the possible methodologies to calculate a correction coefficient have been presented. In theory, the coefficient related to the level of connection, C_{con}, can be calculated as:

$$C_{con} = \left(\sum_j L_j \right) / L_{con}$$

where L_{con} is the length of the network obtained by adding to the current configuration some virtual segments, as short as possible, able to link the cycle segments each other and thus to generate a fully connected network of cycling paths.

As it concerns the other coefficients, at present, we only identify the main ones as follows:

k_w width of the lane;
k_s separation from other flows of mobility;
k_p pavement;
k_l lighting;
k_c number and type of crossing.

Additional coefficients and their values need to be defined considering the context conditions of the paths and thus, they require further in-depth studies.

6   CONCLUSIONS

This study, focusing on the Italian experience, showed how it is possible to consider some conditions that can improve or impede the fruition of an urban cycle path, making it real or illusory urban infrastructure for the sustainable mobility (Festa, 2018). The anomalies that hinder good fruition of a cycle path must be considered as dyscrasias that can be measured and reduced through adequate interventions. Even though more in-depth analyses are also needed to validate the methodological procedure, at present, a first result is a comparison among several anomalies that can occur and the individuation of possible solutions. This study considers an indispensable methodological approach the integration between the optimization of land use and mobility planning. Further development of the research will be focused on the setting-up of an algorithm useful for estimating the level of dysfunction of an urban cycle path.

REFERENCES

Cole-Hunter,T., Donaire-Gonzalez, D.; Curto, A., et al. (2015). Objective correlates and determinants of bicycle commuting propensity in an urban environment. Transp. Res. Part D Transp. Environ. 2015, 40.

Comune di Napoli, Servizio Mobilità sostenibile (2012). Istituzione di un percorso di mobilità ciclistica "Bagnoli – Piazza Garibaldi".

Festa, D.C., (2018). Nuovo ruolo della mobilità ciclistica nei sistemi di trasporto urbani. In; Giuliani, G., and Maternini, F.(eds)Mobilità ciclistica e sicurezza. Egaf edizioni srl: Forlì, Italy, 2018.

Fiorillo A., Laurenti M., Bono, L.(eds.) (2018). Ecosistema Urbano 2018. Rapporto sulle performance ambientali delle città. Legambiente.

Fistola, R., Gallo, M., La Rocca, R. A., (2012). La mobilità insostenibile: impatti dannosi del traffico veicolare sulla salute umana. In Moccia F. D. (a cura di), Città senza petrolio, Edizioni Scientifiche Italianepagg. 75–87.

ISFORT (2018). 15° Rapporto sulla mobilità degli italiani. Osservatorio Audimob.

Mantuano A., Bernardi S., Rupi F.(2017).Cyclist gaze behavior in urban space: An eye-tracking experiment on the bicycle network of Bologna. Case Studies on Transport Policy 5 (2017) 408–416, a Journal on transport Research Society.

Muñoz, B.; Monzon, A.; López, E., (2016). Transition to a cyclable city: Latent variables a□ecting bicycle commuting. Transport. Res. Part A Policy Pract. 2016, 84.

An operational framework for healthy regeneration practices

T. Congiu, A. Plaisant & S. Unali
Department of Architecture, Design and Urban Planning, The University of Sassari (Sardinia), Italy

ABSTRACT: In contemporary society, health promotion extends to daily dimension of the individual and the environment. As a result urban planning programmes and networks involving cities worldwide, such as the Healthy Cities Network program are embracing the principles of public health into their goals.

The perception of a good quality of life is closely related to the interaction between people and space. Thus, the design of the living environment in its various aspects (Burton&Mitchell 2006, Edwards &Tsouros, 2008) becomes critical and needs to be oriented (Capolongo et al. 2018).

The aim of the paper is to provide a reference framework with guidelines for decision-makers with which to inform and evaluate public policies and practices of urban regeneration programs with respect to urban health principles. The implementation of the operational framework in a peripheral district in Cagliari (Italy) is proposed as an example.

1   CONCEPTUAL FRAMEWORK

The relationship between city and health establish new roles and connections among three fundamental dimensions: health, environment and urban living, whose research fields currently move sectorially, without interactions. The definition of project requirements for planning healthy living environments moves from these reciprocal influences. Some conceptual models can support in this task.

The health dimension focuses on the determinants of health, classified according to 4 main levels of influence (Barton & Grant, 2006; Dahlgreen &Whitehead, 2006) and 6 provisions from the Chronic Care Model (Coleman et al., 2009; Epping-Jordan et al., 2004). The operational outcomes of this new perspective on planning and design action is the incorporation of people's habits as well as of factors and qualities of the built environment with an influence on public health.

The environmental dimension assumes ecosystem services as models for cities and peri-urban areas regeneration, distinguished in 4 categories (Millennium Ecosystem Assessment, 2003) and defined by 3 conditions (Boyd & Banzhalf, 2005). Therefore, by designing and implementing green infrastructures and its integrated functions (Allen, 2012; Socco, 2008), in terms of attractive sites, green ways, information and service nodes and areas of landscape significance, we contribute to provide a wide range of ecosystem services, some of which are cultural, visual, aesthetic and social, which support human wellbeing and the physical, psychological and social health of communities.

The urban living dimension refers to the Urban Health Planning model, specifically to the requirements of the City Resilience Index (Arup, 2012) and the new social housing frontiers. The CRI measures the resilience of cities, as their ability to function, to intercept the needs of vulnerable groups and to ensure the degree of health protection to citizens and places.

The urban regeneration experiences of Community Hub (2016) is worthy of remark with respect to new housing models which perform social cohesion and community welfare through collaborative forms of management of collective services and spaces (Nava, 2016; Voci, 2015).

2 MATERIALS AND METHODS

Besides the conceptual models described in section 1, additional methodological orientations for urban planning and design can be drawn by some real experiences and practices in different sectors and geographic contexts. Together, these two sources of knowledge supported the definition of a comprehensive reference framework of project requirements for health-oriented urban planning.

Starting from the needs of vulnerable categories, Burton&Mitchell (2006) identifies 6 accessibility urban design requirements for a dementia friendly city: familiarity, legibility, distinctive character, accessibility, comfort and security.

By arguing the concept of Adapted Physical Activity (APA) Van Coppenolle (1981) analysed the environmental factors that influence the participation of people with vulnerabilities in PA and subdivided them into barriers and facilitators (Leonardi, 2004).

A similar classification of the built environment characteristics affecting the usability of the city for people with special needs is proposed by Cecchini et al. (2018) who defined a set of spatial requirements for promoting the quality of urban life of people with Autism Spectrum Disorder (ASD). In particular the provision of micro "quiet spaces" along local streets together with traffic calming measures to reduce congestion and noise can support ASD people in the autonomous accomplishment of their daily routine.

Another important operational support is offered by ICT and telematic tools, especially those based on user-generated contents. Their extended availability facilitates daily life, for example by supporting the management of mobility across cities and regions as well as the promotion and use of places and services (info and sharing mobility, e-ticketing, booking services, etc.). Some planning tools based on open source geo database, allow to generate itineraries according to the preferences and abilities of users (Sorrel, 2013) as well as inform about the accessibility to services and points of interest for specific types of users (http://www.torino bebi.it). The common denominator of these experiences is the intention to promote active lifestyles and achieve better physical, mental and social health status by enhancing the access to a variety of information and services.

3 CASE STUDY

Conceptual models and real planning experiences supported us in the identification of project requirements to include in healthy urban regeneration processes. The resulting requirements were then better specified according to the characters of the context of implementation. This phase led to a system of project coordinates intended as a list of strategic contextualised objectives to attain by mean of specific actions. In order to support planners and designers in the definition of interventions an operational framework articulated into types of actions, operational objectives and context-oriented project attentions was developed (Table 1).

We tested this method in the redevelopment project of Sant'Avendrace district in Cagliari (Italy) launched during the 2016 national Extraordinary Program of intervention for urban renewal of peripheral areas[1].

In Sant'Avendrace the redesign of public spaces and urban services as a system of multifunctional and multidimensional connections contributed to improve public health, environmental quality and urban liveability.

More precisely, project actions can be grouped as follows:

1. provision of new collective spaces and services that promote active living made accessible at both local and metropolitan level;

---

1. Extraordinary Program of intervention for urban redevelopment and security of peripheral areas", Law n. 218 of 2015, Italy.

Table 1. The operational framework (example for one type of action).

People and environment health-oriented requirement	Reference model	Type of action	Project attentions
Life support Cultural and recreational services Ecosystem approach Multifunctional network Attractive sites Landscape value Access routes	**Ecosystem services** **Green infra-structures**	**c1)** design new spaces and services, according to an ecosystemic and ecological approach	**c1.1) low environmental impact** solutions with innovative, environmentally friendly and recycled materials with ecological certification. **c1.2) access through alternative integrated transport modes** (pedestrian and cycle paths, tpl, sharing mobility,...) **c1.3) integrated public/private open spaces.** **c1.4) lighting design** with low energy equipment
Healthy homes and environments Safe pedestrian and cycling networks Integration between public services and spaces	**SUSTAINABLE MOBILITY AND URBAN RESILIENCE**	**c2)** design physical and visual connections towards -cultural and landscape elements	**c2.1) urban vegetation** for filtering effect and micro-climate mitigation, urban drainage, noise reduction, acoustic and visual comfort **c2.2) combination of species** based on natural dynamics of vegetation. Use indigenous species, and limit aloft species. **c2.3) soil permeability and drainage systems** compatible with paths in terms of safety, comfort, attractiveness and quality of the equipment.

2. construction and management of green infrastructures, intended as functional services to improve citizens' health and well-being;
3. implementation of a new housing model based on the determinants of well-being with emphasis for collaborative forms of spaces and services management as effective driver for community building and empowerment;
4. improvement of physical and functional accessibility to urban opportunities for making the city more usable for all categories of citizens.

4 CONCLUSIONS

The relationship between city and health, often sector-based needs to be tackled with a multidimensional approach.

The proposed operational framework allows to unveil the urban capabilities of the study context and the existent spatial and social drivers to activate in order to create an effective healthy environment for human life.

In the case study a system of connective and accessible public spaces brings together existing urban functions and newly designed services and land uses leading to a living environment consistent with the principles of healthy cities and encouraging an unconventional organization of urban services corresponding to a better quality of life.

REFERENCES

Allen, W. L., (2012), Advancing green infrastructure at all scales: from landscape to site, Environmental Practice 14 (1):17–25.

ARUP (2012) City Resilience Index. Understanding and measuring city resilience. International Development & The Rockefeller Foundation.

Barton, H, Grant, M, (2006). A health map for the local human habitat, Journal of the Royal Society for the Promotion of Public Health, 126(6):252–261.

Boyd, J, and Banzhaf, S, (2006). What Are Ecosystem Services? The Need for Standardized Environmental Accounting Units. Ecological Economics. 63. 616–626.

Burton, E, Mitchell, L, (2006). Inclusive Urban Design. Streets for Life. Elsevier. Oxford.

Capolongo, S, Rebecchi, A, Dettori, M, et al. (2018). Healthy Design and Urban Planning Strategies, Actions, and Policy to Achieve Salutogenic Cities. Int. J. Environ. Res. Public Health, 15:2698.

Cecchini, A, Congiu, T, Talu, V, Tola, G (2018). Mobility policies and extra-small projects for improving mobility of people with Autism Spectrum Disorder.Sustainability, 10(9):3256.

Coleman, K, Austin, BT, Brach. C, Wagner, EH (2009). Evidence on the Chronic Care Model in the new millennium. Health Aff Proj Hope, 28(1):75–85.

Dahlgreen, G, Whitehead, M, (2006). Levelling up (part 2): a discussion paper on European strategies for tackling social inequities in health, WHO Europe, Copenhagen, p.19.

Edwards, P, Tsouros, A D, (2008). A healthy city is an active city: a physical activity planning guide. WHO Regional Office for Europe. Copenhagen.

Epping-Jordan JE, Pruitt SD, et al. Epping-Jordan, J, Pruitt, S, Bengoa, R, and Wagner, E, (2004). Improving the quality of health care for chronic conditions, in "Qual Saf Health Care", 13, pp.: 299–305.

Leonardi, M (a cura di) (2004) ICF Classificazione Internazionale del Funzionamento, della Disabilità e della Salute. Edizioni Erickson. Gardolo (TN).

Millennium Ecosystem Assessment (2003) Ecosystems and Human Well-being. A Report of the Conceptual Framework Working Group of the Millennium Ecosystem Assessment. Island Press. Washington.

Nava, E et al. (2016) Community Hub. I luoghi puri impazziscono. http://www.communityhub.it/wpcon tent/uploads/2016/10/Community-Hub.compressed.pdf.

Socco, C.C, Cavaliere, A., Guarini, S. M,. (2008). L'infrastruttura verde come sistema di reti. Working paper P04/2008. OCS - Dipartimento Interateneo Territorio - Politecnico e Università di Torino.

Sorrel, C (2015). How Komoot Built The Best Bike-Route Mapping App, Fast Company 16/ 11/2015. https://www.fastcompany.com/3052410/how-komoot-built-the-best-bike-route-mapping-app.

Van Coppenolle, H (). ADAPT, Programma Europeo di Attività Fisica Adattata. Source: http://www.kuleuven.be/thenapa/pdfs/adapt1/italy.pdf.

Voci, M. C. (2015) Il «social housing 2.0» aumenta l'offerta di servizi, in "Il Sole 24 Ore Casa 24", 12/11/ novembre 202015. Source: http://www.ilsole24ore.com/art/casa/2015-11-12/il-social-housing-20-aumenta-l-offerta-servizi-122737.shtml.

Conclusive remarks

Pedestrians, Urban Spaces and Health – Tira, Pezzagno & Richiedei (eds)
© 2020 Taylor & Francis Group, London, ISBN 978-0-367-46171-3

Scientific outputs and research needs through bibliometric mapping of LWC discussion

A. Richiedei & M. Pezzagno
Department of Civil Engineering, Architecture, Land, Environment and Mathematics (DICATAM), University of Brescia, Italy

ABSTRACT: The results of a conference, in general, are presented in a round table discussion at the end of the work where the various chairmen summarize the topics covered in their sessions. The purpose of this paper is to bring an objective analysis of the most recurrent and investigated topics from the papers written by the researchers for the Conference "Living and Walking in Cities" (LWC2019) and presented in the proccedings publication book. The keywords of the book as a whole and for the three mainstreams that characterized the Conference are analysed with the method proposed by the software VOSviewer (Van Eck & Waltman, 2011). The aim is to identify the most interesting clusters, peculiarities, strengths and the possible future developments for the Conference.

1 INTRODUCTION. THE "LIVING AND WALKING IN CITIES 2019" CONFERENCE

To analyze the topics covered by the XXIV International Conference "Living and Walking in Cities 2019" (LWC2019), it is necessary to briefly describe its structure. The goal of this event is to gather researchers, road users, administrators, technicians, city representatives and experts to discuss problems that affect the safety of pedestrians in the city, especially of children and persons with reduced mobility. The conference attracts practitioners and researchers who can find detailed presentations on policy issues, best practices and research findings across the broad spectrum of urban and transport planning. The conference covers international issues, national and local policies and project implementation at the local level.

Planners and practitioners are being asked to improve and restructure towns, transportation infrastructure and public spaces. They are finding solutions for resilience in the face of threats posed by climate change, energy and infrastructure security. At the same time, they need to develop hard and soft measures to improve the safety of walking and cycling which affects health and fitness.

The LWC2019 Conference took place in Brescia on 12th and 13th September 2019 and was structured in 3 Plenary mainstream sessions and 9 Parallel Workshops. The Plenary sessions had only one physical outcome: the written contribution of Prof. Cantisani, while of the 47 papers presented in the Parallel Workshops, 45 were collected in the Conference proccedings.

The Conference was structured in 3 mainstreams evoked by the title of the LWC2019 Conference "Pedestrians, Urban Spaces and Health" and illustrated by the abstracts/papers received:

1. Walking experiences
2. Urban spaces and Redevelopment
3. Healthy cities (as Urban resilience and for Frail users).

The titles of Parallel Workshops presentations corresponding to the Plenary Mainstream topics are presented in Table 1.

Tab. 1.　Titles of mainstream and parallel workshops of the LWC2019 Conference.

LWC2019 Plenary Mainstreams	Parallel Workshops
Walking experiences	• Network and infrastructure to improve pedestrian mobility
	• Green infrastructures/Public transport
	• Soft mobility and perception of the urban landscape
Urban spaces and Redevelopment	• Walkability and Redevelopment
	• Pleasant and attractive Public spaces
	• Promoting Healthy Cities
Healthy cities as Urban resilience	• Sustainable and resilient Urban spaces
Healthy cities for Weakest users	• Pedestrian Road Safety
	• Healthy Cities for All

2　OBJECTIVES AND METHODOLOGICAL APPROACH

The main goal of this paper is to analyse, from the lexical point of view, the Conference proccedings and identify the recurrent and most used keywords in order to understand the phenomena and problems specifically addressed in the scientific debate carried out during the Conference. It is also possible to identify the links between the keywords and therefore between the topics addressed to highlight the underlying connections and to promot a greater interdisciplinarity of the research.

The analysis is carried out using VOSviewer software: automatic term identification for bibliometric mapping. Van Eck et al. explain briefly that «given a corpus of documents, we first identify the main topics in the corpus. This is done using a technique called probabilistic latent semantic analysis (Hofmann 2001). Given the main topics, we then identify in the corpus the words and phrases that are strongly associated with only one or only a few topics. These words and phrases are selected as the terms to be included in a term map» *(Van Eck et al., 2010)*. Using this software it is possible to realize distance-base maps of items (keywords) of a text. «Distance-based maps are maps in which the distance between two items reflects the strength of the relation between the items. A smaller distance generally indicates a stronger relation» (Van Eck and Waltman, 2010). In distance-based maps, it easy to identify related item clusters.

The VOSviewer software permits the creatation of different maps through the following views:

• *Density view.* «Each point in a map has a colour that depends on the density of items at that point. That is, the colour of a point in a map depends on the number of items in the neighbourhood of the point and on the importance of the neighbouring items. The density view is particularly useful to get an overview of the general structure of a map and to draw attention to the most important areas in a map» (Van Eck and Waltman, 2010);
• *Cluster density view.* «This view is available only if items have been assigned to clusters. The cluster density view is similar to the ordinary density view except that the density of items is displayed separately for each cluster of items. The cluster density view is particularly useful to get an overview of the assignment of items to clusters and of the way in which clusters of items are related to each other» (Van Eck and Waltman, 2010).

It also possible a *Label view* and a *Scatter view,* but these are not used in this analysis.

The result of this analysis is a Density map and a Cluster density map for the Proceedings keywords and for the three mainstream keywords. The keywords, for clarity, are indicated in quotes (e.g. 'keyword').

3　RESEARCH RESULT: LWC2019 BIBLIOMETRIC MAPPING

After cleaning the text of the so-called "empty words" for this research (such as 'university', 'study', 'aim', 'number', 'paper', 'figure', etc.) and from the names of the cities in the case studies that are necessarily repeated several times in the papers, VOSviewer produced the maps of the items/keywords of the book shown in Figure 1 and Figure 2.

The most used keywords (repeated more than 60 times) are: 1) 'planning', 2) 'development', 3) 'pedestrian', 4) 'path', 5) 'safety', 6) 'life' and 7) 'policy'.

In Figure 2 the clusters are identified in red, green and blu. Even without being able to make changes in the software clustering modes, it is possible to notice, within a margin of approximation, that the words relating to the field of "pedestrians" or "street" are indicated

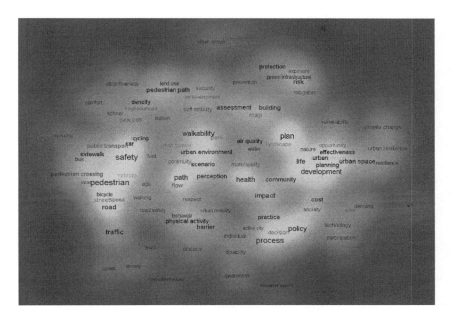

Figure 1.   Density map of the Proceedings keywords (elaboration of VOSviewer software).

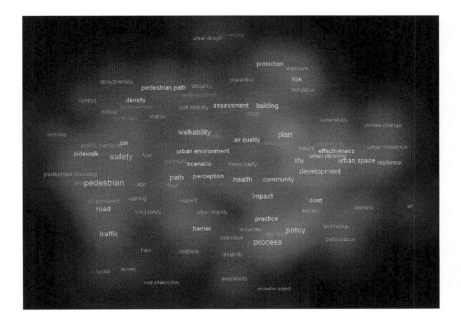

Figure 2.   Cluster density map of the Proceedings keywords (elaboration of VOSviewer software).

in red clouds, the words related to "health" or "life" are indicated in blue clouds and ones linked to the "urban space" or "urban planning" are indicated in green clouds.

The most recurring keywords (repeated more than 40 times) should also be organised in "meaning group": research objects ('pedestrian', 'territory', 'community', 'street/road' and 'life'); research methods ('framework', 'measure', 'map', 'scenario', 'flow' and 'difference'); tools ('planning', 'urban planning' and 'policy') threats ('barrier', 'cost' and 'risk') and opportunities ('safety' and 'sustainability').

By mapping the keywords of the papers grouped according to the corresponding mainstream it is possible to deepen the understanding of specific issues.

Figures 3, 4 and 5 illustrate the density map of the 3 conference mainstreams: Walking experiences (Figure 3), Urban spaces and Redevelopment (Figure 4) and Healthy cities (Figura 5).

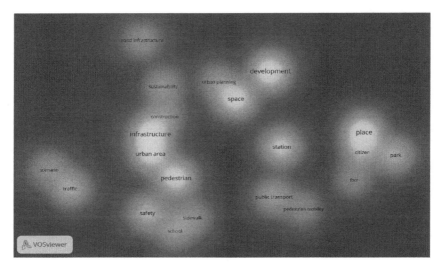

Figure 3. Keywords density map of the mainstream Walking experiences (elaboration of VOSviewer software).

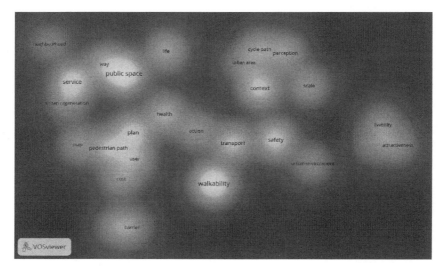

Figure 4. Keywords density map of the mainstream Urban spaces and Redevelopment (elaboration of VOSviewer software).

262

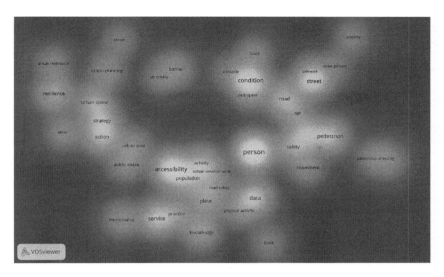

Figure 5. Keywords density map of the mainstream Healthy cities (elaboration of VOSviewer software).

The most used (repeated more than 20 times) keywords in the mainstream Walking experiences are: 1) 'place'/'space', 2) 'development', 3) 'infrastructure', 4) 'pedestrian', 5) 'urban area', 6) 'safety', 7) 'traffic' and 9) 'park'.

The most used (repeated more than 20 times) keywords in the mainstream Urban spaces and Redevelopment are: 1) 'walkability', 2) 'public space', 3) 'service', 4) 'safety', 5) 'context', 6) 'plan', 7) 'transport', 8) 'pedestrian path' and 9) 'health'.

The most used (repeated more than 25 times) keywords in the mainstream Healthy cities are: 1) 'person', 2) 'accessibility', 3) 'condition', 4) 'data', 5) 'service', 6) 'pedestrian', 7) 'action', 8) 'behaviour' and 9) 'resilience'. In this mainstream, the frequency of words occurrence is more relevant than in the others.

4 DISCUSSION AND CONCLUSIONS

Some keywords are obvious (pedestrian, urban space or place) while some others give food for thought.

The first two keywords of the proceedings book are easily linkable: 'planning' and 'development'. The spatial planning to improve urban development is a topical issue - a goal to be achieved - and starting from it there are many research insights that can be explored: case studies, best practices, policy or governance activities, concrete actions or strategies, etc. It can be considered a very fruitful theme.

The keyword 'safety' is recurrent in all three mainstreams. It is mentioned several times in Urban Spaces and Redevelopment. This theme has been significant and constantly present in LWC conferences since the first edition in 1994. The safety of the urban environment and road safety for all users, with greater emphasis on the more frail users (elderly, children, disabled, etc.), is a key and feature theme for the researchers of Town and Regional Planning and Transport group at the University of Brescia (and not only).

The theme of 'services' is also recurrent for the mainstream Urban spaces and Redevelopment and Healthy cities probably precisely because open public spaces (including streets, squares and pedestrian paths) represent a real service for the community and their use, on foot or by bicycle, is also closely related to community health. Health has been defined by the WHO as a state of complete physical, mental and social well-being and not the mere absence of disease (WHO, 1984), while more recently health has been defined as the ability to adapt and manage oneself in the face of social challenges, both physical and emotional (Huber et al., 2011). One of the determining aspects of health is, therefore, the autonomy of the individual which can also be expressed in the ability to move on foot safely in walkable and accessible spaces.

'Walkability' and 'accessibility' are often regarded as synonyms and are strongly present respectively in the mainstream of Urban spaces and Redevelopment and Healthy cities and emphasize the previously mentioned strong bond.

In the item maps, there are also some "peripheral" themes that are present, but not excessively stressed in the papers. These themes represent the cross-contamination between different disciplines and the research opportunities that should be strengthened. One of these themes is the 'risk', accompanied by the keywords 'protection', 'prevention' and 'mitigation'. It may have its own development autonomy in research on cities, but it can also be oriented towards the theme of 'climate change' and 'resilience'.

Another issue concerns the quality of the urban spaces as emphasised by the words 'comfort', 'attractiveness', 'livability' but also 'air quality' and 'park'. It is very versatile and can be approached in completely different ways depending on the point of view of the researcher or the reference mainstream/framework. Last but not least the issue of 'participation' which is accompanied, with originality, by the 'awareness'. Both represent a key element for the success of public 'policies' but at the same time, they are also a 'barrier'.

ACKNOWLEDGMENTS

We thanks Alastair Tweedie for text review.

REFERENCES

Van Eck, N.J., & Waltman, L. (2011). Text mining and visualization using VOSviewer. ISSI Newsletter, 7(3), 50–54.
Van Eck, N.J., Waltman, L., Noyons, E.C.M. et al. Automatic term identification for bibliometric mapping. Scientometrics 82, 581–596 (2010). https://doi.org/10.1007/s11192-010-0173-0.
Hofmann, T. (2001). Unsupervised learning by probabilistic latent semantic analysis. Machine Learning, 42(1–2), 177–196.
Van Eck, N.J., Waltman, L.. Software survey: VOSviewer, a computer program for bibliometric mapping. Scientometrics 84, 523–538 (2010). https://doi.org/10.1007/s11192-009-0146-3.
WHO - Word Health Organization (1948). Constitution of -Word Health Organization (WHO). Geneva (World Basic Documents).
Hube Machteld, Knottnerus J. André, Green Lawrence, van der Horst Henriëtte, Jadad Alejandro R., Kromhout Daan, Leonard Brian, Lorig Kate, Loureiro Maria Isabel, van der Meer Jos W. M., Schnabel Paul, Smith Richard, van Weel Chri and Smid Henk (2011). How should we define health? BMJ, DOI:10.1136/bmj.d4163.

Author index